From The Telegraph
To The Internet

From The Telegraph To The Internet

by

Morton Bahr

President
COMMUNICATIONS WORKERS OF AMERICA

NATIONAL PRESS BOOKS, INC.

————

Layout for this book was produced exclusively by graphic designers at The Kamber Group who are members of the Communications Workers of America.

Library of Congress Cataloging-in-Publication Data

Bahr, Morton
From The Telegraph To The Internet
By Morton Bahr — 1st Edition
Foreword by Senator Edward M. Kennedy
324 pp., 156 x 22.5 cm
ISBN 1-56649-949-6
1. Labor unions — United States
2. Continuing education

PRINTED IN THE UNITED STATES OF AMERICA

Dedication

To my wonderful family …

Florence Bahr
My wife and best friend throughout the years
covered by this book.

My children, Dan and Janice.

And My Granddaughters
Heather, Shelley, Nikki and Alison

It is for them and their generation that the Trade Union
Movement continues the struggle to make the
American Dream a reality.

I love you all.

In Memoriam

Joe Volpe, Mike Mignon, Sal Vitacco, Mickey Gill
and Bob Murphy.

Without them, there would have been no beginning.

Acknowledgments

Without the assistance of those listed below, and so many others, this book would still be a dream.

STEPHEN WEEKS — For his patience in keeping me focussed and his ability to get me to remember experiences long forgotten.

MICHAEL GRACE — For freely giving of his time to collaborate with me in organizing and sharpening this work.

FLORENCE BAHR — For the countless hours spent editing each chapter with special emphasis on its timeliness.

DR. JULES PAGANO — For his friendship, encouragement and for sharing his remembrances of the very early days of CWA.

YVETTE TAYLOR — For her typing and retyping these chapters, always with a smile.

CINDY OLNEY — My secretary who always does those many little things that enable me to get through the busiest of days.

DR. MICHAEL MACCOBY — Who reminded me that I had moved from a 2 to a 10 (on a scale of ten) in favor of a new progressive labor-management relationship in America. (Truth be known, I occasionally drop sharply from the ten.)

STEVE BEVIS — My editor-in-chief who made it all come together.

T his book is written in commemoration of the 60th Anniversary of the Communications Workers of America (CWA).

Collectively, this is the story of how CWA has repeatedly confronted technological and workplace change, grown dramatically and in the process made life better for hundreds of thousands of workers and their families.

Most importantly, this is a story of, for and about the women and men who are CWA—past, present and future ... a story of the commitment, the dedication, the hopes, the dreams, the courage and the accomplishments of CWA members throughout North America.

Contents

Foreword

BY UNITED STATES SENATOR EDWARD M. KENNEDY
OF MASSACHUSETTS

In many respects, the one constant in the 60-year history of the Communications Workers of America (CWA) has been change. From the formative days of the telephone industry to today's burgeoning information industry, the members of CWA have experienced technological and workplace change as few others have. The manner in which individuals and organizations respond to the challenge of change speaks volumes about their collective character, commitment and ultimate success.

In *From The Telegraph To The Internet*, CWA President Morty Bahr tells the story of the Communications Workers of America as he has lived it. And Morty Bahr has lived and breathed CWA since the day in 1951 when he became an in-plant organizer for the union at Mackay Radio in New York City.

I have known Morty Bahr for more than two decades. I have been to his home. I have been with his family. I have seen him around his grandchildren. I have seen him with his CWA family on countless occasions. I have also seen him with Presidents, Senators, Representatives and corporate leaders from throughout America. This much I know—Morty Bahr knows what matters to working people and he cares deeply about workers and their families. His actions and accomplishments over the past five decades are a testament to that commitment.

The idea of "making life better" is a theme that I have often heard when I've been around Morty Bahr. Not surprisingly, this is a theme that is interwoven throughout *From The Telegraph To The Internet*.

This book chronicles one of the true success stories of the American labor movement. Throughout its history, CWA has continually found itself confronted with workplace and societal change. CWA has repeatedly responded by developing innovative programs and strategies that have resulted in a better life for its members and countless other workers around the world. During a period when some have tried to argue that the labor movement is no longer appropriate in today's increasingly high-tech workplace, CWA has effectively shattered that theory by adding thousands of new members to the union in recent years.

Today's CWA membership reflects the convergence of workers and industries that characterizes the new information age. In addition to telecommunications members, CWA now represents workers in printing, publishing and newspapers, in broadcast and cable, in law enforcement, as well as thousands of workers in the public sector, including health care and university professionals.

While *From The Telegraph To The Internet* is written from the unique and personal perspective of someone who has been on the front lines during so many of CWA's momentous events and achievements, perhaps its most important message comes in the form of providing a candid picture of what the successful 21st century union will look like.

This book spans an extraordinary period of change and progress in America's history and provides a wonderful, personal account of the struggles and successes of the Communications

Workers of America. It is written with the compassion and spirit that have long characterized Morty Bahr.

All of us who know Morty are proud of his leadership and grateful for his friendship. By putting his own special perspective on the history of the Communications Workers of America, Morty Bahr has provided all of us who recognize and appreciate the countless accomplishments of CWA and the American labor movement with an outstanding gift.

CHAPTER ONE

The Labor Movement– Crisis or Opportunity?

My first real union experience has stayed with me for my entire life. During World War II, I was an eighteen-year-old radio officer aboard the Merchant Marine Liberty Ship *SS Thorstein Veblen*. The ship's master was Angus McIntyre, a grizzled sea veteran who was not much admired or liked by the crew. Fortunately, he was not typical of the many fine officers and seamen in the U.S. Merchant Marine.

This was my first ship and my first voyage. A couple of weeks out at sea, the Chief Mate invited the off-duty officers into his office to discuss the negotiations that the unions were conducting with the War Shipping Administration. We were members of three different unions – the Masters, Mates and Pilots, the Marine Engineers and my union, the American Communications Association.

Unbeknown to us, Captain McIntyre was hiding outside the door, listening to every word that we were saying. After a few minutes, he kicked in the door in a wild rage, shouting, "There will be no union talk aboard my vessel!"

For the entire time that I sailed under him, there was no open union talk aboard his ship.

Times haven't changed much. Just ask any worker in a non-union workplace who has tried to form a union. Capt. McIntyre may be long dead, but his hatred of unions is shared by many managers in America today.

Is it any wonder why organized labor faces an uncertain future?

Today, fewer than 10 percent of the private sector workforce belongs to labor unions. As total union membership declined, organized labor saw its power diminish, thus lessening our ability to be effective advocates for working families.

Instead of moving forward, we are often fighting to just maintain the status quo. Labor organizers are finding it increasingly difficult to bring new workers into the labor movement in face of the sophisticated and expensive campaigns waged by companies to create fear in the workplace and to keep the union out.

At the bargaining table, unions often have to fight just to keep existing jobs, wages and benefits at their present levels.

In the political arena, our efforts, with few exceptions, have been to block legislation that would hurt working families and to fight to retain legislative gains won over the past sixty years—rather than advancing the agenda of working families.

In part, unions are victims of their own success. In the 1930s and 1940s, unions had great success organizing the mass production industries, particularly auto, steel, and rubber. The unions grew as the industries expanded to supply the Allied war effort and kept expanding after the war as we helped rebuild Europe and Japan.

During the postwar period, the telephone system also experienced rapid growth, and with it, the numbers of unionized workers in the Bell System and the other smaller telephone companies also increased.

The unions in these basic industries, directly employing more than three million workers, became extraordinarily efficient in servicing their members. Organizing largely was automatic—as the companies grew and hired more workers, unions grew proportionately. Since the Communications Workers of America (CWA) was the union that represented most of the telephone workers, our union grew along with the Bell System.

In the telephone industry, organizing essentially took place when Western Electric built a new factory. CWA and IBEW would fight fiercely for the right to represent the workers. The results, after many years of wasted resources, were about 50-50. CWA always had a long-term objective of winning over the thousands of telephone workers represented by unions that decided to stay independent after CWA was formed.

Labor's long run of success came to an end in the 1970s. America's dominance over the world economy—the legacy of our victory in World War II—was no longer unchallenged. Other nations began to successfully compete against U.S. companies. German and Japanese steelmills, for example, produced and exported low-cost steel to the U.S., which substantially cut into American production. The decline of American manufacturing, a development long in coming, was not recognized by many until it was already too late. Americans were shocked to read that the vast World Trade Center in New York City was built without one ounce of U.S. steel, a realization that brought home to many the fact that the era of American industrial dominance had ended.

How did it happen? Technology certainly played a role. U.S. trade policy also opened our markets to foreign imports but failed to require that our trading partners open their markets to us in return. In the auto industry, this failed trade policy, coupled with domestic car makers' slow response to the desires of American consumers, resulted in a dramatic loss of market share in a relatively short period of time. Soon the effects spread to the rubber industry, steel and many other sectors of the economy.

Competition also began to enter the telephone industry in the 1970s. Telephone and other equipment manufacturers competed with AT&T's Western Electric manufacturing plants. MCI established itself as a major long distance competitor to AT&T and began to cream-skim (attracting lucrative business contracts with wholesale-like prices while ignoring less profitable services to small businesses and residential consumers). Meanwhile, regulatory agencies prevented AT&T from competing in the marketplace on an equal footing. While the prices that AT&T and its operating companies charged were regulated, their competitors were free to charge whatever it took to get the customers. Hundreds of small companies sprang up to compete with the employers of CWA workers in installing and repairing phone equipment in businesses and homes.

Still, we were relatively unaffected by the economic and labor difficulties of the 1970s. Not only the automobile and related industries suffered decline during this period, but change occurred throughout the labor movement in railroads, construction, airlines, and many other unionized industries. Union membership declined dramatically. The International Association of Machinists once proudly boasted of a million members in a sign that hung over their office building in downtown Washington. By the end of the decade, the sign had to be taken down—a victim of membership losses.

Labor unions were faced with a choice—adapt or die. In order to remain viable, unions had to change their organizing strategies. No longer could we simply grow along with our industries. We now had to reach out and organize new workers, either within our traditional industries or outside of them. It was not a matter of choosing between servicing our existing members or organizing new workers. Instead, it was a matter of survival. Unions had to bring in new members to ensure that we would still be around to serve our existing members.

Lynn Williams, former president of United Steelworkers of America, told me his union had to decide virtually overnight to become an organizing union in order to reverse the decline while continuing to provide quality service to their members. Change, he said, was expedited by pain: the steelworkers lost almost half their membership and had to lay off a proportionate number of their staff.

During the 1970s, CWA's membership, for the most part, was concentrated in the heavily regulated telephone industry, so we were able to avoid the precipitous decline that other unions were experiencing. But we could see the handwriting on the wall. The world around us was changing rapidly. Changing times called for changing strategies.

CWA leadership developed a variety of programs, both within the union and in partnership with our employers, to prepare for the changes brought about by the technological development and global competition that we knew were inevitable. We also realized that some type of competition would be introduced into the telephone industry, although we never envisioned such a radical break-up of the Bell monopoly as eventually unfolded. Technological change and global competition now present even more challenges for CWA and for all of organized labor as we stand on the threshold of a new century.

Does crisis or opportunity await us? The Chinese symbol for crisis and opportunity is the same. Emulating the wisdom of the ancient Chinese, we choose to view today's challenges as a great opportunity for CWA members and the American labor movement.

Only 10 percent of the private sector workforce in America belongs to labor unions. Is that a crisis or opportunity? Instead of bemoaning the fact that only 10 percent of the private sector workforce is organized, I look at it this way—90 percent are out there ready to be organized.

I believe we can seize upon the opportunities created by historic change and help bring about a resurgence of the American labor movement. CWA is uniquely situated to take a leadership role in moving the labor movement into the 21st century and in making the American dream available again for workers. But we must first clearly articulate the new role of trade unions in a rapidly changing world and determine how unions can best serve the needs of workers in a global economy.

Young people entering the workforce, most of whom were brought up in nonunion households in an era of increasing union hostility, must learn about the history and role of organized labor in our economy. Union leaders have always lamented that the press almost never reports the accomplishments that organized labor contributes to the overall good of the community. Labor agreements rarely get reported. Strikes always do. The many contributions that union members make to their communities are largely ignored. The few corrupt union leaders are loudly, and often justifiably, criticized.

I don't look for this to change. In fact, as control of the media is concentrated more and more in the hands of fewer and fewer owners, negative portrayals of the labor movement are likely to increase.

This book is an attempt to correct the record, and tell the whole story about the future of the labor movement from the perspective of one union among many so that all may have a clear and positive vision of the future. I am writing this book on the 60th anniversary of the Communications Workers of America to tell our story, where we came from, where we are now and where we plan to go. By so doing, we hope that you will not only get a sense of appreciation for our union, but also the role we and our members play in building a better America, and a better world. Our mission is simple, yet profound—to make life better ... for our members and their families, for our communities, and for all of society.

CWA is a progressive, dynamic and forward-thinking union. We began in 1938 as a union for Bell Telephone system employees. Now we represent more than 630,000 workers in telecommunications, printing and newspapers, radio and television broadcasting, cable television, public service, public safety, higher education, the airline industry and other technical, professional and administrative employees in both the public and private sectors.

As the broad range of our membership indicates, CWA is no longer only a telephone union. We have created a new type of union, not based within a single industry, but representing workers from different industries, all of whom have common concerns, both in the workplace and in the communities in which they live.

Together we are building *the* union for the 21st century.

CWA is a union with strong values and great commitment to our members, their families, and the communities in which they live and work. Our members aren't greedy, but they want decent wages. They never want to have to choose between their jobs and caring for a sick child or parent. They want to be certain their pensions and health care will be protected when they retire.

Our members want dignity on the job and a voice in their workplaces. As we organize more professional and technical workers, we find that this highly skilled and committed workforce also wants a greater role in the decision-making process where they work. They look to our union and the collective bargaining process as a way to bring about greater democracy in their workplace. They want their union to play a role in their career development by negotiating education and training programs.

We are poised to bring about dramatic and lasting change, not only in the telecommunications industry, but throughout the global economy. In part, CWA has the opportunity to be an effective agent of that change, ensuring that the information revolution is humane, equitable and democratic; that working families are allowed to share in the wealth they help create, and that their rights and interests are protected.

When he addressed the CWA convention in 1994, Vice President Gore charged us with a special responsibility. He said that since CWA members understood the power of the new technology, we had to play an important role in helping form public policy to be sure that all citizens share equitably in the changes that are transforming America and the world.

CWA members have been able to maintain a high standard of living. They enjoy higher wages and more benefits than workers who do not belong to unions. In general, union workers are much better off than nonunion workers. Union wages are on average 20 percent higher than nonunion wages. Union workers enjoy significantly better benefits and retirement packages.

They have better job security and recourse to address injustices in the workplace.

But doing better than the other guy is not good enough. We can't view our union or our members in isolation. I see organized labor as a broad economic and social movement. We don't just represent workers in the workplace and forget about them when they go home; our job doesn't end at the plant gates. Every day of the year, in a variety of ways, in thousands of communities around the country, CWA members make life better for all working families.

American labor unions have traditionally been deeply involved in their communities. The forty-hour workweek, overtime pay, employer-provided health care, and a variety of government programs from Social Security to health and safety regulations, unemployment insurance and workers compensation—all were fought for and won by organized labor. And they benefit all working families, not just union members.

We need to return to what made the labor movement great, advancing the cause of working America as a whole. At the same time, we have to look ahead toward a future of dramatic and rapid change.

CWA has always been a union with an eye toward the future. Our founding president, Joe Beirne, had a vision of a national telecommunications union, which was realized when CWA was formed in 1947. Joe's next objective was to win national bargaining with the Bell System. The fight for national bargaining took CWA nearly thirty years, but it was finally achieved in 1974. The progressive leadership of our second president, Glenn Watts, helped fulfill Joe Beirne's vision by focusing on the future, not just of CWA, but the workforce and the economy as a whole.

CWA leaders always looked to the future because we had to.

Our members work in industries on the cutting edge of technological change. We have already experienced the wrenching changes of automation, corporate shakeups, reorganizations, mergers, downsizing, and other competitive pressures that are now coming to bear upon the rest of the economy.

As the union representing 85 percent of the workers in the Bell System, CWA faced tremendous challenges following the Bell System breakup on January 1, 1984. An industry that had been a monopoly for nearly a century was suddenly dismantled and subjected to fierce competition. Workers who had previously felt very secure about their jobs and futures were now faced with new employers, relocations, layoffs, and countless other traumas. As a union, we went from one national bargaining table in 1983 to about forty-eight in 1986. And as competitive pressures mounted, bargaining became a much more complex process. We were impacted, for the first time, by differences in regional economies, as the regional operating companies were not able to absorb regional downturns the way the national Bell System historically had.

In the years after divestiture, more than 100,000 jobs were lost—most of them at AT&T, although the Bell operating companies also reduced the workforce largely through attrition. While downsizing is always wrenching, CWA was ready for change. We targeted all new lines of business in the telecom companies and expanded our organizing within the new broader industry. However, most of our new growth came from outside the core telecommunications business. This enabled us to maintain a fairly stable membership level, although many local unions felt the pain.

As an employee of Mackay Radio & Telegraph Company, a subsidiary of American Cable and Radio Corporation, a telegraph company, my workplace was the first non-telephone bargaining unit in CWA when we won the NLRB election and joined the union in 1954. That was the beginning of change in CWA, from a telephone union to one that would represent workers in the broad information industry and beyond.

Coming from another field gave me a perspective that I know proved invaluable in my years as vice president and president. I saw that we needed to look beyond our traditional constituencies. While continuing to meet the needs of our current members, we had to bring in new workers. It was up to union leadership to demonstrate to our existing members that reaching out to unorganized workers was in everybody's best interest.

Joe Beirne saw that in order to be effective, the union had to succeed on three separate levels: Collective Bargaining, Organizing, and Community and Political Action. Joe dubbed this concept the Triple Threat, and it continues to be a driving force behind our union today, although we now call it the Triangle.

Collective bargaining is the fundamental role of the union—representing the workers at the bargaining table and protecting their interests with their employers.

Organizing is essential to union growth, to increase the power of working families and reversing the decline of labor. It gives workers empowerment through the solidarity of numbers.

Community and political action are the realms in which the union must exercise and expand its influence. Joe Beirne called CWA "the community-minded union." Our members live in communities all across America—we have members in all 435 congressional districts—and we exercise our responsibility to be active in those communities. Future success will depend on unions playing a larger role in the communities where members live and work. And the union has to be politically involved at all levels—local, state, federal and international—representing the interests of working families and society as a whole.

The three sides of the triangle support each other. Each must be strong for the union to be effective. If we don't organize, we don't grow and we ultimately lose members. This affects our strength at the bargaining table and our presence in the community. If we don't bargain effectively for our members, we can't improve their living standards. That affects our ability to organize and reduces our clout in the political arena and our voice in the community. If we are not active in the community, and if we can't make our voice heard at all levels of government, then we can no longer serve as a voice for working families.

When organizing new members, particularly those who don't come from a union background, we often have to battle misperceptions about organized labor. Most of these misperceptions are the result of antiunion propaganda. But we can't fight a public relations war with corporate America on their terms. The advertising budget of AT&T alone is $1 billion a year. We simply don't have those kinds of resources. What we do have, however, are members who live in every community across the country and the power of their collective actions.

After the 1986 round of collective bargaining, the first one after divestiture, membership participation had to be dramatically increased if we were to be able to properly represent our members in the future. The days of passive membership were over. We needed our members to be more actively involved in the union, the workplace and the community. So we began to develop the practice of mobilization—actions that demonstrate and reinforce solidarity, and express concerns based upon union and human values. In a mobilized union we could no longer tolerate an attitude of "let the other guy do it." Now everyone was "the other guy."

An enormous change took place in CWA between 1986 and 1989. Mobilization became more than a tactical response to a specific problem. Rather, it was the beginning of grass roots involvement at all levels of the union. Most importantly, mobilization became a tool we used in collective bargaining. At times, it served as an alternative to a strike and a way to express solidarity by allowing our workers to remain on the job and simultaneously exert sufficient pressure on their employers to force a settlement.

As we prepared for the 1998 round of telecommunications bargaining, mobilization was largely institutionalized and the results showed in our successes at the bargaining table. Mobilization supports the Triangle in many ways. It encourages education, participation and communication throughout the union. A local union that has a mobilization strategy is a strong and effective voice for justice on the job and in the community. An educated and active rank and file become great salespeople for their union and organized labor in general. They are the front line trade unionists who, when asked, respond with telephone calls and e-mail to members of Congress on legislative issues, talk to reporters, and write letters to the editor.

While education is a major ingredient of any mobilization program, it is much more. Lifelong learning is a primary focus of CWA's mission and a cause that I am personally dedicated to, as a lifelong learner myself. Well before I became president of CWA, it became apparent to me that union representatives needed to continue their education. Our industries, collective bargaining and the world in general were growing more complex. For many years, a union leader's street smarts were a match for management. But our street smarts are no longer enough in today's world.

I came to this conclusion during my years as vice president of District One, responsible for all CWA programs and activities in New Jersey, New York, and New England. I had dropped out of Brooklyn College during World War II. Not only did I go back to school to get my degree, but I encouraged staff and local leaders and members to do the same. I know how difficult it is to juggle work, family and school responsibilities. But I also know how important it is to commit to a lifetime of learning, not only to advance your career, but to develop as a person.

I do all I can to support and encourage lifelong learning among CWA members and for all Americans. I am proud that CWA members take advantage of a wide variety of education programs. Every year, CWA has among the highest participation in the George Meany College Degree Program.

Education and lifelong learning are crucial to building a world class workforce. I don't accept the prediction of many academics that the average American will work for seven different employers during his or her career. But I do know that no longer can a high school graduate go to the telephone company and work at the same job or in the same location until retirement. Now and in the future, workers will work at many different jobs—hopefully with the same employer—as long as they have the educational opportunities to learn new skills.

With the rapid escalation of technology, global competition and mobile capital, how can a union guarantee its members job security? I don't believe we can. But we can offer them employment security.

Employment security is the cornerstone of CWA's collective

bargaining strategy. We have negotiated outstanding jointly administered training and education programs, paid for by the employers, for the vast majority of our members. Through these opportunities, CWA members have the ability to make themselves more employable with their current employer, or, if they choose, prepare for a second career. In any event, they are clearly in charge of their own lives after taking advantage of continuing education opportunities. Lifelong learning benefits employers as well as workers. An educated workforce is a more productive workforce.

We know that without profitable companies in a productive economy, you cannot have good, high paying jobs. Our employers in the telecommunications industry have begun to realize the value of a union workforce. More enlightened CEOs recognize the union's need to grow along with their company. They have responded by agreeing to management noninterference and card check recognition or uncontested elections in the non-represented sectors of their companies and subsidiaries. This means management will permit their employees to decide whether they want a union in the workplace, free from fear, and will recognize the union when a majority of workers indicate support of the union by signing cards or through an election supervised by a third party netural within two weeks after the union asks for the election.

In addition to bringing stability to the company and growth to the union, wall-to-wall representation is a vital job security protection for our members. As technology and the marketplace continue to change the nature of our industries, wider union representation means more job options for the employees.

CWA will play a prominent role in shaping the information industry of the future. That industry will be characterized by competition and convergence. As the marketplace becomes even more globalized and capital becomes even more mobile, competitive pressures will increase. Competition will always tend to favor low-cost producers, unless worker organizations and global solidarity exist to raise the cost of abusive practices. We need to help more American workers organize into unions of their choice, so that the nonunion workforce is not used to pressure us to surrender at the bargaining table. We also need to strengthen coalitions with labor unions in other countries, and help support organizing efforts all over the world.

Convergence is being driven by new technologies, competition and corporate mergers. The telecommunications, publishing, broadcasting, entertainment and computing industries are becoming one massive industry of information services. One can see it in the mergers and acquisitions among the already large and powerful media and telecommunications companies. This process will continue until there are only a handful of major players in every market.

As corporations converge, so must organized labor. Driven by the need to recoup their huge investments, employers will strive to compete by lowering labor costs. A powerful voice representing workers throughout the new information services industry must be present to counter the tremendous concentration of wealth and power of these new multimedia multinationals.

CWA has a parallel mission. We must be prepared to deal with these powerful companies by effectively representing our members at the bargaining table.

Of equal importance, we must be committed strategically and financially to organize the hundreds of thousands of new workers who will be employed. We are poised to make this happen.

It is also up to labor to ensure that the changes brought about by convergence benefit rather than hurt working families. If organized labor isn't there to protect the interests of working families, no one will.

Among our members and future members, we must create an appreciation for the past and generate enthusiasm for the future. The two are inseparable. Without a strong sense of the challenges we faced in the past and how we succeeded in meeting them, it will be difficult to persuade people that we can successfully confront the challenges of today and tomorrow.

This book is the story of how one union dealt with change over the past 60 years and how it will continue to shape its own future and that of America's working families far into the next century.

CHAPTER TWO

A Life in the Union

I did not come from a union family. My mother and father were immigrants, Russian Jews who came to America as young children at the turn of the century. My father never said to me "Son, you are Democrat and union." Well, we were Democrats, growing up in the New Deal era, huddled around the radio to listen to FDR's fireside chats.

My father was in the retail business most of his life. At one point, he owned a grocery store which went belly-up. During the depression, he traveled around the country looking for work. When times got better, he found a job as a supervisor for the Union News Company, a major newspaper and magazine distribution company and restaurant operator in train terminals.

In 1948, the Union News Company employees in New York went on strike. At the time, I was working two jobs, following a long strike at Mackay Radio. My second job was for the Union News Company in New Jersey—which was not on strike.

On the way to my second job, I passed by a struck news counter. Having just days earlier come off my own picket line, and realizing the person working at that counter was taking the bread away from those fighting for a better contract, I impulsively yelled "scab" as I walked by. Without warning, the strikebreaker leapt over the counter and punched me. Our struggle was short lived and I went on my way.

That evening, my father's boss came to my workplace. He told me that I had jeopardized my father's job by my earlier action. Later, I realized that he didn't threaten my job because I was a union member. My father could have been punished for the "sins" of his son. I had exercised my right of free speech and because a management person disliked what I said, my dad could have been fired.

This experience also has stayed with me my entire life. It showed me how the union protects its members, how unorganized workers are more vulnerable than union workers, and how management often reacts when it feels threatened.

I was born in Brooklyn, as was my wife, Florence, and eventually, our two children, Dan and Janice. I was very young when my family moved to the Bronx, a few blocks from the famous Bronx Zoo, the first zoo where the people were caged and the animals ran free.

I grew up in the tenements. We were poor, as were the vast majority of people living through the depression. You didn't get a new pair of sneakers until two toes came through, and often times you would just put a piece of cardboard in your shoe to cover up the hole.

My mother's brother, who was a bookmaker, lived with us. Business was good for a bookie in those days because people who don't have much money tend to gamble. The local tailor used to give me a nickel each time I brought in one of my uncle's suits. I always remembered years later, how that suit was important enough to that tailor for him to give me a nickel. It was my first lesson in marketing.

At the time, I didn't realize how difficult things were. Relatives generally lived in close proximity to one another and shared what they had. Most of my mother's family lived within walking distance. Only years later, in conversations with my relatives, did I finally understand how tough life really was back then.

In 1938, when I was twelve, we moved back to Brooklyn. The country was coming out of the depression and people were working again. I went to junior high and high school in Brooklyn, graduating from high school a month before my sixteenth birthday.

I was consumed with baseball, playing it in high school and then in a sandlot league. As players began to be drafted into the armed services, the Brooklyn Dodgers organization canvassed the high schools and I was invited to a tryout with the team as a catcher. Even though I didn't make it, the tryout was an experience I would never forget. Imagine what it was like for a sixteen year old kid from Brooklyn and die-hard Dodgers fan, being in the Dodgers' locker room and walking by lockers that bore the names of Carl Furillo, Dolph Camilli, Dixie Walker, Whitlow Wyatt and Cookie Lavegetto. As the kids say today, it was awesome. Years later, in talking with Sandy Koufax about my experience, he said that it was important for me to remember that I had been asked. He was so right.

I went to Brooklyn College and made the varsity baseball team. I had the good fortune to play against Ralph Branca when he pitched for New York University—and I got a hit although they beat us. Branca went on to be a star pitcher for the Dodgers. Unfortunately, he is remembered for serving up the home run ball to Bobby Thompson that gave the hated New York Giants the National League pennant in 1951.

After one year, I left Brooklyn College. America had been at war for two years and victory was not yet in sight. I enlisted in the U.S. Merchant Marine as a radio officer.

I went to a radio school run by the Coast Guard on the lower east side of Manhattan, where I learned to type and became proficient in the Morse Code. I finally was prepared to take the FCC examination for a Second Class Telegraph license. With a good deal of luck, I passed.

I joined Local 2 of the American Communications Association, CIO, the union that represented shipboard radio officers. And, one week past my 18th birthday, I was on my first ship, assigned to the *Liberty Ship SS Thorstein Veblen* which carried 8,000 tons of high octane aviation fuel. We were part of a convoy of 68 ships, all carrying explosive material.

The union dispatcher told me to report to the ship Saturday afternoon. It was docked in Brooklyn. I went aboard and met the chief mate. He showed me where my room was and told me I didn't have to be back until Monday morning.

I was glad that the entire crew was ashore since I had difficulty finding my way to the dock. Having had a year of college, however, I realized that if I followed the ship's rail, I would eventually come to the gangway.

When the ship finally left dockside, I was in a state of blissful ignorance. Since I had my own room as radio officer, I had no knowledge about the habits of other crew members. When I went to bed, I put on my pajamas and generally slept soundly. That is, until perhaps thirty days at sea when the alarm bells went off. Submarines had been sighted on the other side of the convoy. Depth charges were being dropped and guns were being fired. When all hands were called on deck, I learned that most of the crew slept in their clothes with life jackets very handy.

That's when I began to worry.

I had never been on anything but the Staten Island ferry before. The *Veblen* was filled with colorful personalities. The Chief Mate's last ship had been a sailing vessel and the other radio operator (there were two of us) was a professional hobo.

It was a real learning experience for a kid from Brooklyn. After the incident I related at the beginning of this book, when Capt. Angus McIntyre kicked the door in during a union discussion, I learned that my captain was antiunion and that aboard ship the captain is king. McIntyre made life on the ship very difficult—and taught me how important it was for workers not only to be organized, but for their union to have real power. On the *Veblen*, we were organized, but isolated on a ship with an autocratic captain, we had no power to change things.

While the union is powerless at sea where the Captain reigns, grievances were corrected once the ship came home. For example, just months after the war ended in 1945, my ship, the *SS Abraham Clark* docked in Bremerhaven, Germany. A former German U-boat captain was in charge of the prisoners of war assigned to unload the *Clark*. To my chagrin, one day I walked into the officers' dining room and found the U-boat captain having lunch with our captain at his table.

I immediately lodged a verbal protest which the captain was going to ignore. But I had confidence in my union and told him that the first place I go when we get to New York was to the union to report this travesty. Within five minutes, the German POW was back on deck.

After the war ended, I had stayed on in the Merchant Marine. I had met Florence Slobodow in June 1943 on a blind date. It was love at first sight—at least for me. Her father was a cooper—he made barrels—and a member of the Teamsters. In fact, he started on a horse and wagon. Florence was a union member where she worked managing a cleaning store.

I saw Florence whenever my ship came back to the U.S. In early June of 1945, I had returned from a voyage to India. The ship was berthed in Newport News, Virginia. The war in Europe was over and the U.S. was beginning to send relief cargoes to Europe.

After being home a couple of weeks, I said goodbye to Florence and returned to the ship. We went from Newport News to Boca Grande, Florida, where we took on 2,000 tons of phosphate fertilizer, then proceeded to Galveston, where we loaded tons of food staples and then headed for Marseilles.

Several days out in the Atlantic, we discovered we were taking on water. The captain told me to radio the company, Grace Line, and ask for instructions. We were told to make New York if we could. If not, head for the closest port.

We made it to New York. Divers were sent down and soon discovered that we had been loaded with too much cargo, which caused the ship to hit the bottom and break some hull plates. What apparently saved us from a disaster was that when the phosphate fertilizer on the ship's bottom got wet, it formed a cement-like substance and slowed the seepage of water into the ship.

So, there I was, unexpectedly home in New York. It was a Saturday and Florence was working. I walked in and surprised her, and we went out and got married the next day. Our families were both away at the time. When they found out we had gotten married, our parents were none too happy at first. We were teenagers and ours was a marriage that statistics say doesn't last. But now we've been married fifty-three years, with two kids and four grandchildren—all union. I believe this marriage will last.

Florence was pregnant and due to give birth in October 1946. I spoke to her in August before leaving Marseilles. She had been to the doctor that day and was told everything was fine. So, imagine my surprise when the marine ship-to-shore station on Long Island told me they had a message for the ship and it was addressed to me. I froze, translated the morse code in my head and learned that I had become a father. Three days later, when the party ended, I called the station and asked for a repeat of the radiogram so I could copy it for the record.

We were in the Bay of Biscay on our way from France to Germany when my son Dan was born. I didn't get home until Dan was six weeks old.

I received my discharge in November 1946 at the end of what was my last voyage after 29 months at sea. In the Merchant Marine I had been earning a good salary, plus room and board. It was a good life—in peace time, with little responsibility. I now had a family and didn't know anything else. So, I told Florence I would make one more voyage.

For someone who had been alone most of her pregnancy as well as through delivery, her response was understandable. She said: "Go ahead. But the baby and I won't be here when you return."

Florence had already anticipated my concerns and fears. She presented me with classified ads for radio operator jobs at Mackay Radio and RCA Communication. I was hired by Mackay Radio (later to become American Cable & Radio and then ITT World Communications) on March 11, 1947. Had Florence not had the foresight she displayed, my career—and, indeed, our lives—would have taken a completely different turn.

I became a member of American Communications Association Local 10. Mackay was a closed shop, meaning you had to be a member of the union before you got a job. In effect, the union hired you. (This was before the Taft-Hartley Act, which made closed shops illegal.) Even before I took the test, the shop chairman, Harley Benson, told me I had the job. It paid $49 dollars a week.

As I look back, our union at Mackay was very powerful. To some degree, more powerful than the management in the workplace. So powerful, in fact, that it caused the company to plan to bust us. When our contract expired on New Year's Day, 1948, we went on strike. The company forced the strike and had a strategy to win.

The strike lasted 90 days. It was one of the coldest and snowiest winters on record. I began picket duty midnight to 2:00 a.m. but as the strike dragged on and the picket lines became thinner, by the end of the strike, I picketed midnight to 6:00 a.m.

Throughout that difficult time, with a small baby and no income, Florence and I never considered my crossing the picket line. Honoring the picket line was a basic union principle that we took for granted. It needed no discussion.

Unfortunately, by going on strike, we played right into the company's hands. This was the first major strike since the Taft-Hartley Act had been passed over President Truman's veto in the summer of 1947. Among other things, Taft-Hartley required union officers to sign a noncommunist affidavit before they could use the facilities of the National Labor Relations Board. The officers of our union either wouldn't or couldn't sign the affidavit.

Management was aware of this, and ready to take advantage. About six weeks into the strike, the company went to the Labor Board and filed a petition for decertification. Since the union couldn't appear on the ballot, we were decertified. Notwithstanding, we remained on strike seven more weeks.

When we returned to work I found out what it was like to be without a union. Union leaders told us to return to work on April 1, reporting to the shifts we were on prior to the strike. I was working the all-night shift and was the first to leave the job, so I was among the first to return.

We were intentionally humiliated. We were told to line up in the lobby at 67 Broad Street, New York City. There we filled out forms and were sent home with: "Don't call us. We may call you." I wasn't called until April 5.

Mackay Radio had been the company involved in the U.S. Supreme Court case in 1938 where the Court ruled that companies could permanently replace striking workers. Notwithstanding, although scabs were brought in from their Latin American operations, Mackay management never threatened us with permanent replacements. This was also true in later strikes as well.

The reason is simple. Until President Reagan discharged the air traffic controllers in 1981, it was not acceptable corporate conduct to permanently replace strikers. President Reagan's action led to a destabilization of collective bargaining.

Without a union contract, there is no justice. I learned that first hand in the late 1940s. Because we didn't have a union, we worked from 1948 to 1950 with virtually no wage increases. I recall giving Florence my paycheck, her looking at it and handing it back to me, saying: "It's not enough." A paycheck that did not include overtime did not stretch far enough to pay for the necessities of life.

Remember, this was a time when most workers were getting big raises. At Mackay, not only were our wages kept down, but the workplace became a horror. Our work rate was speeded up, and increased pressure was put on us by the supervisors. The company decided that seniority would not be used in the selection of vacations. Someone could have thirty years of service and not get a summer vacation. We lost our ability to bid on shift assignments, and were put on a weekly shift rotation. None of us liked our change in circumstances, but it made us more determined to fight back.

One night, although there were four of us with no work assignment, hour after hour the supervisor gave me excess traffic to copy. I was being set up because I had become a leader—and I bit. The supervisor was trying to provoke me, and unfortunately I did just what he wanted.

I blew my top and was suspended for two days.

I didn't want Florence to have more to worry about than she already did, so I didn't tell her about the incident. She was supportive of my union activity, but I was concerned she would worry that my job was at risk. Since I was working midnight to 7:00 a.m., I had to do something to occupy the time, so I spent the two nights in Times Square movie theaters which, in those days, ran wholesome Hollywood movies all night long. And since I didn't earn much, it was easy to borrow the two days pay.

In December, 1950, a CWA organizer named Joe Volpe was giving out handbills in front of the Western Union office, which was a block away from where I worked. At the time, Western Union in New York was represented by ACA, and CWA was trying to organize them. One of my coworkers, Danny Scarafile, saw Joe and asked him to come down the street and talk to us.

We liked what Joe Volpe had to say about his union. At that time, I and a number of others who worked for Mackay Radio continued to pay union dues to the ACA. With CWA's entry into the picture, we took the next step. We asked for a meeting with the ACA officers. We stressed the importance of their compliance with the Taft-Hartley law and told them that if they chose not to sign the non-communist affidavit, we would go to CWA.

We waited for several hours while the ACA Executive Board met. They finally decided not to sign the affidavit. We left and immediately announced our support for CWA.

Early in 1951, I became an in-plant organizer for CWA. It took three years to get the plant organized, not because the workers didn't want a union, but rather because three different unions were involved and the company used every tactic to delay the election.

Our victory and affiliation with CWA was a significant turning point for both me and the union. At the time, CWA was exclusively a telephone union and our unit was the first non-telephone group to join them.

When CWA arrived on the Mackay property, it came with an interesting history. In the first decades of the 20th century, very few telephone workers were union members. Those who were organized belonged to the International Brotherhood of Electrical Workers (IBEW). During World War I, a presidential order placed the telephone and telegraph system under the control of the federal government, overseen by Postmaster General Albert Burelson. In 1919, Burelson was faced with a strike by the IBEW that virtually tied up phone service in New England and threatened to do the same nationwide. In an attempt to end the strike, Burelson issued a government bulletin acknowledging the right of workers to bargain through committees "chosen by them, to act for them."

Fearing that Burelson's order would stimulate union organizing throughout the company, AT&T established the American Bell Association, which they hoped would supplant the growing independent unions. At that time, companies were forming their own unions to prevent their workers from joining real unions. Company unions would hold meetings on company property, often closely supervised. Dues were very low, and the union officers were paid by the company. As a result, the company unions did little but pay lip service to representing their workers. At the same time, the pressures on the workers to join were immense. By creating company unions, and "encouraging" workers to join, AT&T effectively pushed the IBEW out of the industry.

Company unions dominated the telephone industry until 1935, when Congress passed the National Labor Relations Act (also known as the Wagner Act). The Wagner Act, which remains the cornerstone of labor law today, stated that workers had a right to join a union, and established legal mechanisms to help protect that right. Most importantly, the preamble to the Act, still operative, reads, "It shall be the policy of the United States to foster collective bargaining."

The Wagner Act prohibited company unions, protected union organizing, picketing, strikes and other activities, and established the National Labor Relations Board as the agency to deal with labor disputes. When the Wagner Act was upheld by the Supreme Court in 1937, the company unions in AT&T were effectively busted.

In doing research for this book, I was fascinated to learn that Joe Beirne, CWA's founding president, never aspired to be a union leader. In fact, at one time early in his career, he was a management employee of Western Electric. Joe planned to be a lawyer involved in politics and government.

At the time company unions were banned, Joe was enrolled in college, taking courses at night that would have led to a law degree. He knew that building a union completely independent of the company was essential. His natural leadership qualities came to the surface and were recognized by his coworkers. His prelaw studies made him the one person qualified to write a constitution. The rest was history. Joe Beirne went on to become a great labor leader. He travelled all across the country on busses with his family in tow, organizing his fellow workers.

In 1938, 145,000 telephone workers organized into the National Federation of Telephone Workers. But the NFTW was not a national union, only a federation of local independent unions. This put the NFTW at a disadvantage in negotiating with AT&T. Joe Beirne, now the NFTW's president, wanted a union with national bargaining power with the Bell monopoly. In 1945, he negotiated the first national agreement with the president of AT&T, but thirty-four of the fifty-one affiliated unions broke away and signed separate agreements.

Two years later, AT&T flatly refused to bargain on an industry-wide basis. It tried to divide and conquer the NFTW unions. The NFTW went on strike, but its lack of centralized power caused the strike, and the federation, to fail. Joe Beirne later said, "We were trying to make a federation of unions do the job which only can be done by one union in the telephone industry."

Out of the dissolution of the NFTW, CWA was born. The first convention was held in Miami in the summer of 1947, with two hundred delegates representing 162,000 workers. Joe Beirne was elected president, and he began building one national telecommunications union. In 1949, CWA became affiliated with the Congress of Industrial Organizations (CIO) over the vigorous opposition of AT&T, whose management still had visions of dominating the union by keeping it isolated and outside the mainstream of organized labor.

CWA had an uphill struggle to win the organizing campaign at Mackay, which had become American Cable & Radio. CWA supporters had to overcome the argument that CWA was a telephone union and knew nothing about the telegraph business. We had to convince our coworkers that we would not be swallowed up in a large union that represented so many workers in another field.

We told American Cable and Radio workers that CWA was born out of struggle and strife. It had a dynamic and inspirational leader in Joe Beirne. It bargained with the largest corporation in America and therefore should be more than a match for our company. It had members everywhere and was well respected across the country. Perhaps what was most important—and can serve as a lesson for all organizing campaigns—the CWA in-plant leaders were highly respected among their peers. In April 1954, CWA eventually won in a runoff election against the ACA.

A couple of Sundays later, a mass membership meeting was held in old Webster Hall in Manhattan, a place where a lot of labor activity took place in those days. Even though I was not a candidate, I was unanimously elected temporary local president. The regular elections were held that fall, at which time I was elected for a two-year term, again unopposed.

I was flattered at being recognized by my coworkers in such a way. But, like so many labor leaders before and after, I asked, "what do I do now?" I admit to being frightened. Although I had been active in bringing CWA into our workplace, I had no idea how to organize, much less run, a local union.

In 1954, I attended my first CWA Convention, as a guest. I

was never so proud as when President Beirne introduced me to the convention to the applause of hundreds of delegates.

The next year I attended the CWA Convention in St. Louis as a delegate. The following dialogue from the verbatim minutes is illustrative of the battle we fought to get into CWA as well as the change that needed to occur within CWA to make it the active and growing union it is today.

DELEGATE MORTON BAHR (LOCAL 1172): I would like to ask Jack (Vice President Moran) a question. Your report did not touch on the telegraph end of the communications industry. I would like to know if it is the intention of the organizing department to pursue as vigorously the organization of all radio and cable companies, such as RCA Communications, Western Union Cables into CWA, as it is to organize the Bell System in CWA.

VICE PRESIDENT MORAN: The resolutions which have been passed by this and other conventions in the past have reiterated that our job is to organize all of the communications workers in the U.S. and Canada. And that includes the telegraph, the independents and anything we consider as communications.

DELEGATE BAHR: The reason I raised this point is because my fellow delegate from AC&R and I have been sitting here all week, and we are the only two people in this convention who are not telephone people. And you know, Jack, during the long organizing campaign, that was the thing that was being used against us—that we would be swallowed up in telephone. I don't think so. In fact, I know we won't. But we continually hear reference to this being a telephone union. I would like to suggest that the

Chair request everyone here to refer to this union as a Communications Union. *(Applause)*

Following the convention, President Beirne sent a reminder to everyone. CWA was a communications union, not just a telephone union.

Once we joined CWA, our working conditions improved enormously. One incident in particular shows how much our status in the workplace changed. In 1955, when I was local president and still working for the company, the chief shop steward, Sal Vitacco, filed a complaint against the same supervisor who had me suspended some four years earlier for my union activity. The assistant superintendent, Red Pascucci, called me into his office. I was unaware at that time what was transpiring. In his office were the steward and the supervisor. Shortly after I entered, Pascucci told the supervisor: "The next time the union has you on the carpet, just clean out your locker. You're fired."

As the other two departed. Pascucci held me back. "Satisfied?," he asked. He obviously remembered what had happened previously.

"Yes," I said. "Now do I get my two days pay?"

"Be honest," he said, "didn't I just give you more satisfaction than two days pay?"

As I reflected back at this incident many years later, it wasn't the two days pay or the fact that I had been vindicated that really mattered. Instead, it was the clear recognition of our union by the management and a measure of the credibility we had built up in so short a time. It also demonstrated to me that without a union it is virtually impossible to achieve justice on the job.

I was reelected local president in 1956, running without opposition. In December 1957, Ray Hackney, CWA's Vice President in charge of organizing, asked me if I wanted to work for the national union as an organizer in charge of the New York Telephone project.

I was flabbergasted. I didn't believe that anyone in Washington knew me. Besides, I didn't know much about the telephone industry or running an organizing campaign.

The job wasn't permanent, so I decided to give it a shot. What I didn't know at the time was that CWA had had a presence in New York Telephone since 1952, with many different organizers, including several loaned to CWA by the CIO. I was being put there to maintain CWA's interest, but no one thought we would go to an election. Fortunately, no one told me about this because from the first day on the job, I set my sights on bringing this important group of workers into our union.

So, I went from being the president of a local to head of a campaign to bring 18,000 telephone workers into the union. I had never even been inside a telephone building and had no idea what the workers did there.

I didn't know it at the time, but I was about to embark on an adventure that has lasted to this day. From that moment on, my future would be forever tied to CWA.

CHAPTER THREE

Building the Union

Organizing the New York Telephone workers became an all consuming job. I resigned as local president in July 1958, and took a one-year union leave of absence to devote all my energies to the task. When I went in to tell the assistant superintendent, he told me that if I stayed I'd probably become a supervisor. But I've never regretted my decision to leave and go full time for the union.

Prior to this campaign, my organizing experience had been limited to working as a rank and file committee member, bringing my own workplace into the union. Although I knew nothing about the telephone industry, I quickly realized that it was more important for me to learn how to motivate people and build an organization with them, rather than learning everything about their jobs. I spent the first few weeks getting to know the members of our in-plant committee. They were a wonderful group of dedicated men and women who never gave up on CWA.

I asked Vice President Hackney for assistance and specifically requested Joe Volpe, the organizer who brought me into CWA. After a few months, it became clear to us that we really did not know our strength. We determined to go for an election even though we knew we could not win.

The election took place in 1959 and we surprised ourselves

with the strength we showed. The final tally—out of 19,000 total votes—was UTO 10,558, CWA 4,752, and IBEW 1,589.

After that election, we knew that we could eventually win. The campaign for victory was to begin the next day. At least, I thought it would.

I prepared a budget request and sent it to Vice President Hackney. I was told that President Beirne wanted me to personally come before the Executive Board to make the case for my request. Not only was this never done before, but I wasn't even a regular employee of the union at the time. To say I was nervous and apprehensive would be an understatement. But it forced me to be prepared.

We created a battle plan that listed every one of the hundreds of work locations in the downstate New York area and the names of the in-plant committee person or persons at those locations. We had early response machinery to react to breaking events. Nothing was left to chance.

At the time, CWA had an educational institute in Front Royal, Virginia. The Board meeting was held there. I made my presentation, answered some questions and, I believe, shared my enthusiasm.

The Board voted to give me the requested budget. On leaving the meeting, Joe Beirne said, "I'll bet you a case of wine, we don't win."

"I know we will win and I never bet on sure things." I responded.

That was the only time I remember saying something to Joe Beirne when he didn't have a quick comeback.

Joe Beirne liked to challenge his union leaders. In 1959 CWA's Collective Bargaining Policy Committee was meeting in New York

City. I was assigned to work the microphones. During a reception, President Beirne backed me into a wall and poked his finger in my chest.

"Why don't you want to help us build a union in New Jersey?" he asked. There was an opening for a permanent representative position in New Jersey. I had turned it down because it would have meant leaving the New York Telephone campaign.

Before Joe had a chance to work me over, someone told him that his guest had arrived, Robert Briscoe, The Lord Mayor of Dublin. Joe told Ray Hackney to bring me to Washington the following week for a continued discussion.

Again, this gave me time to prepare.

Joe, Ray Hackney and I met for lunch in the old National Democratic Club in Washington. I impressed upon Joe my commitment to my present assignment and told him that I knew we could win. He finally said: "Okay. But Ray, you are a witness that this man has no commitment for a job. That he can be sent back to work for AC&R any time."

With that, I was even more determined to win the New York Telephone election.

The second election was held in February 1961. We won by 454 votes. IBEW ran third and was eliminated from the runoff that ensued. That brought into CWA more than 18,000 workers in the downstate New York area.

Things were a little different for the 6,000 workers in upstate New York. The independent union was eliminated in the first election and the runoff was between CWA and IBEW. The work of Ray Hackney's assistant and my dear friend, Gus Cramer, was instrumental in the outcome. He successfully persuaded the president of the independent union to write a letter endorsing

CWA in the runoff, and we won by a "landslide" of 24 votes. That letter was the difference.

These two wins, the largest union gains in the private sector in many years, increased our presence in the region and gave President Beirne more strength in dealing with the Bell System.

There is nothing like success to win promotions. Shortly after the elections, I was put on the permanent staff. Several months later, I was appointed to the newly created post of New York State Director.

When I joined the union in 1954, there were perhaps 25,000 members in the District that comprised New York, New Jersey and the five New England states. The membership in New York was small and was virtually all Bell System locals, with the notable exception of Local 1170 that represented the employees at Rochester Telephone Company. The bulk of the membership was in New Jersey.

The District Director (which we now call Vice President) was Mary Hanscom. Mary was one of the early pioneers in telephone unionism, a tough and smart leader. She never asked anyone to do something she wouldn't do herself, including leading strikes that landed her in jail.

The New York Telephone elections dramatically changed the demographics of the district, leading Mary to decide to step down as vice president in 1963 and stay on as New Jersey Director for a while before she retired. Her assistant, George Miller, a native of Kentucky and Tennessee, announced he would run for the post. George was opposed by a local president from the New York Telephone Unit. I strongly supported George. When he won, I became his assistant.

To be a good assistant, you must be able to submerge your

own identity, and just work for your principal. Because George was a hard worker and we were philosophically attuned, I was able to do that for him. Although he did a really good job, he faced tough opposition every two years. At the time, I hadn't realized how very parochial New Yorkers can be — and George was just not one of "us." A very large segment of the New York locals never accepted George.

In 1969, a senior staff person, one whom I had recommended for the position, announced his candidacy against George. I knew it would be an extremely close election and called in every chit ever owed to me. We won by some 700 votes.

This election had potentially serious consequences for me. Had the opposition candidate won, I would have left the district. But since we won, there were positive consequences for me. Several weeks after the election, President Beirne asked George Miller and me to come to his office. He told me that George had agreed to come to Washington as his assistant and that he was polling the Executive Board to name me District Vice President until the next convention when I would have to run.

I was sworn in as Vice President of District One in August 1969 and served in that post for sixteen years.

During my term as vice president I was fortunate enough to never have anyone run against me, which allowed me to make nonpolitical decisions, because unions, like other organizations, are intensely political. Being unopposed, I didn't have to take political consequences into the decision-making process. Because of this, I had the freedom to make the personnel decisions I felt needed to be made for the long-term interests of the union.

As an elected officer, both as Vice President and President, I was blessed in having the most loyal and talented trade unionists as part of my administrative teams. As Vice President I had Gene Mays, Ted Watkins, Jan Pierce, Larry Cohen, Clara Allen, Ed McCann, Larry Mancino, Don Sanchez, Jack O'Brien, and Jean Fawcett. As President, working with me are Dina Beaumont, Ted Watkins (retired), Larry Cohen and Ron Allen. They are all marvelous individuals who were and are totally dedicated to the members of CWA.

During my time as vice president, I was able to make District One leadership more diverse and representative. When I was elected, we had one African American on staff, Gene Mays. I made Gene my assistant. We put the first black female on staff in New Jersey, Gloria Shepperson. The New Jersey director at that time was Don Sanchez, a Hispanic.

Because of our extraordinary growth in New Jersey, I was able to have our staff largely reflect the faces of the membership. We had sixteen staff, eight male and eight female with minorities among them. We had two directors, one female and one male.

So, we were a very diverse district at a time when the rest of the leadership was undergoing change, but still predominantly white and male.

When I came into the union in 1954, we had about 25,000 members in District One. When I left in 1985, we had 120,000. Not only was our staff diverse, but so was our membership. We were so successful in expanding into the public sector that the District was evenly divided between private sector workers and public sector workers.

It was interesting to watch the interaction between the two employee groups at district meetings. Private sector leadership soon learned that revenue at the state and local levels that came in from taxes we paid was just as important to the public workers as a rate increase won by the telephone company was to the telephone workers. There came a slow but firm realization that they were all workers and the name of their employer on the paycheck was irrelevant. All of our members, regardless of what work they did, had the same problems and the same dreams. They wanted to raise their children in safe neighborhoods, see them go to good schools, be able to send them to college while they enjoyed dignity on the job and looked forward to retirement with a good quality of life.

And they knew the union was the vehicle to accomplish it.

Since CWA was born out of the Bell System, it was influenced dramatically by the culture of our dominant employer. The company was large and maternalistic, and fostered the idea that there were two types of people—those who worked for the telephone company and everyone else. People called it having a Bell Head. Not having worked for the phone system, I didn't have a Bell Head and consciously tried hard not to develop one.

In 1968, President Beirne had all of the elected officers and their assistants for a week long session in the Front Royal Institute with a group of psychologists. During one particular exercise dealing with organizing, I observed that if CWA was looking at a non-Bell company that had 10,000 unorganized workers and a group of one hundred at a new Western Electric facility, we would focus on the one hundred, because they "belonged" to us. Although Bell employees were our traditional members, I felt strongly that focusing exclusively on them, and not trying to organize other workers, would be detrimental to the union.

By then we had already started organizing in the public sector. In 1966 a group of New York City workers called the Municipal Management Society (MMS) visited us. They represented thousands of top level city workers who in the private sector would be excluded from union representation because they were considered management. New York City public employees were covered by the "Little Wagner Act" (passed by then mayor Robert Wagner, son of the Senator after whom the national Wagner Act was named). It provided for elections and exclusive representation.

The MMS leaders said that every time they filed for a certification certificate, it was denied on various grounds. I advised them that since we were not a dues collection agency, we would not agree to take them in unless we were able to win recognition.

I went to see the New York City Commissioner of Labor, James McFadden, who is still a good friend of mine. McFadden brought in the Department's counsel, Phil Ruffo. They leveled with me. Every time the MMS filed for certification, a request came from the head of the New York City Central Labor Council, Harry Van Arsdale, Jr., to find a reason to reject it because they were an unaffiliated labor union. And, since so many different reasons were given to block the various requests, it would take an order from the mayor to untangle it.

I passed all this on to President Beirne and he told me to make an appointment for us with Mayor Wagner. We met with the mayor the day Wagner returned from his honeymoon, having married the sister of the fire commissioner. I didn't know at the time that Mayor Wagner's first wife (who had died) was related to Annie Beirne, Joe's wife.

For more than an hour, these two old friends sat around talking about old times. I recall the mayor telling Joe about how he called on Labor Secretary Arthur Goldberg for advice on how to settle the newspaper strike of a couple of years earlier that shut down all of the city's papers. He said that Arthur listened intently and then pensively said: "Bob, you have a hell of a problem."

Finally, Mayor Wagner asked: "What can I do for you, Joe?"

"I don't know." Joe responded. "Tell him Morty."

So I told the Mayor the entire story of the MMS. Wagner got McFadden on the phone and said, "Jim, I have Joe Beirne and Morty Bahr here, and they have a problem. Take care of it right away."

I went immediately to McFadden's office and told him that the Mayor told me to call him if the matter wasn't taken care of within a week.

Of course, it was. The process of certification, signing up new members and recognition moved at a rapid pace. Today, the MMS is CWA Local 1180, with about 8,000 members.

There was a strong lesson to a young trade unionist in this experience. If CWA was not a politically active union, the door to the Mayor's office would never have opened to Joe Beirne and myself.

Joe Beirne was a great union leader, absolutely the president that CWA needed in order to build the union and make it strong. National bargaining with the Bell System was his dream and it took his entire life to see that dream come true.

In January 1974, Joe Beirne was dying of cancer. He got out of his sick bed to announce that the Bell System had agreed to national bargaining. This was a goal Joe had since 1947, and the culmination of three decades of hard work. It was fitting that Joe lived to see this great achievement. Seven months later, Joe would be dead. He died on Labor Day that year.

Because of his declining health, Joe hadn't sought reelection in

1974 and Glenn Watts became the next president. Joe had let it be known that Glenn was his choice. Glenn had risen through the ranks from local president to district director, to assistant to the president, to secretary-treasurer and now was president himself.

Early in 1976, Executive Vice President George Gill died. I received a call from President Watts who told me that his assistant would take over Gill's position and he would like me to come to Washington as his assistant.

I was flattered but not ready to leave what we were building in District One. When I declined the offer, I told Glenn that there had to be a better system. We seemed to wait for someone to die before personnel decisions were made. Then someone has to change their entire life overnight. Glenn agreed with me and said that we would talk some more about it when we got together at the Democratic National Convention in New York in July, 1976.

At the convention, Glenn said he did not know when he would retire but would like me to consider being his successor.

Again, I was flattered and somewhat overwhelmed. But I forgot about the conversation shortly thereafter. We were building something important in District One. I had a team that enjoyed each other, as well as working together and looked forward to spending my entire career there.

In September, 1984, we had our annual unstructured Executive Board meeting in Keystone, Colorado. At the unstructured board meeting, there is no prepared agenda. Each board member lists those matters he or she would like discussed.

Glenn opened the meeting by saying he wanted to list the first item and wrote on the blackboard: "Election of Officers 1985." I commented that the elections weren't until 1986. He responded, no, this is what he wanted on the board.

After some twenty other items were listed, we went back to the first item—1985 elections.

Glenn said that several months earlier, Secretary-Treasurer Louis Knecht came to his house and told him he was going to retire at the 1985 convention, one year before his term expired. Glenn said he then confided in Lou that he was thinking of retiring himself. President Watts said he was driven to the early retirement because he felt it was important to give his successor at least one year to prepare for the first negotiations following the Bell System breakup.

Glenn's decision was a demonstration of union statesmanship. It also set a stage for what in many unions becomes a divisive and even disastrous battle for the two top spots. The closest CWA ever came to a split was when Joe Beirne was challenged by Vice President Tommy Jones back in 1957. After Joe overwhelmingly defeated Jones, he welcomed all of Jones' supporters with open arms and no hard feelings. That too was a demonstration of union statesmanship.

Concerned that a divisive election struggle would hurt the union, Glenn suggested that we arrive at a decision by private consensus. If we were all willing, he would meet with each of us privately over the next three days and then tell us what our consensus was. Everyone agreed.

"So," Glenn said, "who is interested?"

After what seemed like five minutes of silence but probably was only a few seconds, I said: "I am." That triggered Executive Vice President Booe to say he was as well. Glenn did not do the same for the Secretary-Treasurer position.

Over the next three days, in thirty-minute segments, Glenn

met with all members of the Board. On Wednesday afternoon, he called to tell me that I was the candidate. He congratulated me and said he believed the Board acted correctly. When I appeared to be surprised about his personal support, he said, "Why are you surprised? I told you I wanted you to be my successor."

"Yes," I responded, "but that was eight years ago and you never spoke to me in the interim." In fact, in 1980, we had a major disagreement—when I supported Senator Kennedy against President Carter in the primary.

"I never changed my mind," Glenn said.

He asked me not to say anything until he spoke to Jim Booe. That apparently didn't happen until late that evening. Jim called me at 5:00 a.m. to congratulate me and tell me that he did not want to be Secretary-Treasurer. I asked him to meet me in an hour.

I told Jim how fortunate I was throughout my career to be surrounded by people with whom I did not have to worry about turning my back. If he felt he could work with me like that I would like to have him in the number two job. Further, I was aware of certain responsibilities he had that I told him he could keep, such as being a member of the Democratic National Committee. After some discussion, he agreed.

At 8:00 a.m., we told President Watts we had a ticket.

Not only was all of this well received around the country, but the union was more united than ever. Somehow, *The Wall Street Journal* got the story. I got two calls in succession. The first was from Senator Kennedy who congratulated me and then said: "I have just one big complaint. You are becoming president before me."

The next call came from Governor Cuomo. He simply said:

"Don't forget where you came from." And, I never have.

Before I became president, the union had been lucky in that each president was the right person for the right time. Joe Beirne was absolutely the key guy for CWA in its early stages. He was a great organizer, a motivator, a teacher. He was tough; a lot of people were frightened of him. But he built this union from scratch. I don't know anyone else who would have taken his family on a bus and traveled the country to build a union. He continues to be an inspiration to those of us who knew and admired him.

Glenn Watts took Joe Beirne's vision to a new level. He presided over the years of national bargaining, and helped make the union more diverse and more democratic. He also began preparing the union for a future that he knew would be tumultuous. His idea for a Committee on the Future and other forward-looking projects not only helped CWA deal with changes we are now experiencing, but also got the union thinking more progressively. Because of Glenn's foresight, today we are better able to face the challenges brought about by dramatic technological and workplace changes.

When I took over from Glenn, one of the biggest challenges we would ever face was already a reality—the breakup of AT&T.

CHAPTER FOUR

The Bell Breakup

O n January 8, 1982, the telecommunications industry and our union were changed forever. AT&T, the Department of Justice and the White House announced their agreement to a consent decree that settled the government's antitrust suit against the Bell monopoly being tried in Federal Court. This historic agreement would break up the monopoly and forever alter CWA, as well as the relationship all Americans had with their telephone company.

Some months afterwards, during a time when American phone users were in a state of great confusion regarding the changes brought about by divestiture, I was invited to discuss the subject with about a hundred arbitrators, all of them attorneys, at Hofstra University in Long Island, New York.

At the end of my remarks, one of the lawyers asked: "What did the union do to block what happened?"

I asked: "Is there one among you who finds fault with the question?"

The room was silent.

I pointed out that the labor movement had no role in the consent decree. We didn't have any involvement in the process of divestiture until after the agreement was reached and all the major decisions had already been made.

The concern that all of us should have as Americans, I told the Hofstra group, was that the complete restructuring of our telephone industry was engineered by an assistant attorney general hired just for this case, the chairman of the board of AT&T who answered only to his shareholders, and a federal judge appointed for life. Ed Meese, President Reagan's chief of staff, had signed off on the consent decree after a one-hour meeting with Assistant Attorney General William Baxter and AT&T Chairman Charles Brown, notwithstanding the fact that the Secretaries of Defense and Commerce and the Chairman of the Joint Chiefs of Staff all had serious concerns about the breakup. Despite these concerns and others, the Reagan Administration never even held a cabinet meeting concerning the break up of the Bell System.

A competitive telephone industry was a long time in coming. For more than a decade, AT&T had been able to successfully resist, or at least limit, competition that was slowly being introduced. A 1956 consent decree had already restricted AT&T to only providing telephone service and not moving into other lines of business.

It was clear, however, that technology and the growing globalization of the marketplace were making the 1934 Communications Act, which regulated the telephone industry, obsolete. A technological revolution was creating a new information industry and the entry of major new players was inevitable. The question was not whether there would be competition, but how that competition would take shape. It was time for Congress to amend the 1934 Act or draft a new one. A few feeble efforts were made, but they quickly failed.

This set the stage for Ronald Reagan. When Reagan took office in 1981, his advisors saw the breakup of AT&T as an ideological cause and initiated a new and more powerful antitrust action against AT&T. Since Attorney General William French Smith's family had significant holdings of AT&T stock and thus had an obvious conflict of interest, William Baxter was appointed as a special assistant attorney general just to prosecute the suit.

The trial was held in U.S. District Court in Washington, D.C., before District Judge Harold Greene. In his opening remarks to the court, Mr. Baxter said that if the judge found AT&T guilty of violating the Sherman Antitrust Act, the government sought the total breakup of the company as the remedy.

The Administration wanted the component parts of AT&T— the 22 Bell Operating Companies that provided local dial-tone services, Long Lines which handled long distance telephone calls, and Western Electric which manufactured telephones, switches and other equipment—to all be independent companies

It took the government several months to present its case. When the prosecution rested, AT&T lawyers made a standard motion to dismiss. In rejecting AT&T's motion based solely on the evidence produced by the government, Judge Greene implied that he believed the government's case was overwhelming.

Understanding that the judge was unlikely to rule in their favor, and considering the Administration's ambitious break-up plan, AT&T Chairman Charles Brown, along with the company's general counsel, met secretly with Assistant Attorney General Baxter in Vail, Colorado, over the Christmas holidays in 1981, while the Court was in recess.

On January 5 through 7, 1982, CWA leaders and AT&T corporate executives were meeting at Harvard's Kennedy School of Government training for the implementation of a new program called Quality of Work Life, which was designed to empower workers in the workplace.

There was not a single leak about what was happening in Washington. We learned later, however, that the presidents of every Bell Company were secreted away, incommunicado. Without knowing precisely what it would be, these executives were waiting for the word from Chairman Brown.

At 9:00 a.m. on January 8, Glenn Watts called all the CWA officers and told us that an extraordinary announcement would be forthcoming that morning.

Hours later, the agreement on the consent decree was made public. As events unfolded, it became clear that faced with an adverse judicial decision and a complete breakup of the company, Charles Brown went out of his way to get the best deal he could through negotiations.

Clearly, none of the parties to the settlement were concerned with the impact of the consent decree on the employees who built the world's best telecommunications system, nor the subscribers who used it daily. The company seemed only to be looking after the best interests of their stockholders.

In an attempt to get quick judicial approval of the decree, the negotiating parties hastened off to see a judge in a Newark, New Jersey Federal Court, where the 1956 decree had been approved. They contended this was simply a modification of that final judgement, because they wanted to avoid the triggering of the Tunney Act, a requirement for a sixty-day period of public comment before judicial approval is granted.

CWA was one of the first to raise concerns about the decree. Just a few days after the filing in Newark, Judge Greene exerted his jurisdiction and took control of the case. CWA immediately filed for intervenor status and was approved. We were the only union with an official role in the commenting period and thus were able to raise with the judge matters of concern to the workers.

This was the biggest corporate shakeup in history. Yet, it was designed and carried out by a handful of powerful men. The settlement had enormous impact on the country. Telephones had been in wide use for almost 100 years, and 94 percent of all homes had at least one phone.

At the same time, there was no open competition in any other telecommunications system anywhere in the world. With virtually no exceptions, every other country's telephone system was government-owned and operated. The Bell breakup would have a dramatic impact on the rest of the world. Shortly after the American monopoly was broken, the British and Japanese began to change their telecommunications systems. A global tidal wave of privatization quickly followed and continues to this day.

In every country where the telecommunications system was owned by the government, so was the postal service. Invariably, revenues from the telephone side were used to subsidize the postal side. Once competition entered the marketplace, not only did the postal subsidies have to cease, but government needed to find revenue from tax dollars to continually upgrade the telecommunications systems to remain competitive. No country was able to do this. Thus, various forms of privatization took place and are still continuing.

On January 13, 1982, President Watts called a meeting of our Executive Board to review everything we knew about the break up of the Bell System.

We released a statement, along with an analysis that was sent to all CWA local unions. Even then, CWA took a long view of this momentous event and its impact on our members and our union. We were determined to control our own destiny despite the massive changes that were ahead of us.

The Board's statement said, in part:

"[A]midst this far-reaching change—the one constant is CWA. CWA is a union whose democratic traditions and values do not change. Its concerns for the well-being of the members is never-changing, it is undeterred in the goal of having stable, financially sound and progressive Locals which can provide, through this period, the strength and continuity upon which our members rely....

"We will chart our own course."

And, chart our own course we did.

The statement was designed to reassure our members that their union was on top of the situation. We also wanted to assure the local leaders that their local unions would remain intact; that government or company actions would not change their jurisdiction.

We were concerned, from our reading of the decree, because of the restrictions laid out therein, that the operating companies would simply become dial tone companies. We questioned their financial viability. So, for example, we were supportive of Judge Greene "persuading" AT&T to agree to amend the decree to move the lucrative yellow pages businesses to the operating companies.

We called for prompt negotiations with AT&T to ensure that AT&T and present Bell operating companies and successor operating companies recognized CWA and their responsibilities to honor contractual obligations as well as to continue to bargain in good faith on all matters that affect employment of CWA members. We would also be addressing employee job and pension rights.

Since this is not a book about the divestiture of the Bell System, suffice it to say that successorship rights were negotiated. All employees were assured of following their work and without Bell System objection, Congress passed CWA-proposed legislation guaranteeing all employees who were on the payroll December 31, 1983, pension portability when they moved from one former Bell company to another.

When people ask "are labor unions still relevant?," I would ask them this question in response: "Who would have protected the rights of the some half-million Bell System employees were it not for their union?" At no time during the entire process, did any party to the antitrust suit address the concerns and futures of the employees who built the system.

Our January 13, 1982 board meeting was memorable for still another reason. Washington and the Northeast were being hit by a severe snowstorm. At noon, I asked Glenn if he would mind if I left for home; that if I didn't leave now, I would never get out of D.C. So, the last anyone saw of me was getting into a cab about 12:30 p.m. on the way to National Airport.

The airport was a mess. The plows were barely visible on the runways. About an hour later, they announced an effort would be made to get the 3:00 p.m. shuttle to La Guardia off, but first the plane was going to be de-iced. That was enough to send a chill down the spine of this seasoned traveler.

I called my secretary in New York and asked her to get me a reservation on Amtrak and that I would call her when I got to Union Station. When I got to the station, I also called home, but there was no answer. So I boarded the train.

It took more than six hours to reach Penn Station. Before getting on the Long Island Railroad to go home, I called Florence.

She asked: "Where in the world are you?"

I told her that I had taken the train because the airport was closed. I didn't understand why she sounded so concerned. That's when she told me that everyone was calling and looking for me— they thought I might be dead. An Air Florida plane had crashed into one of the bridges on the approach to National Airport and since I had not been heard from, the conclusion was that I was in one of the taxis that was on the bridge at the time.

After the first call, Florence phoned our headquarters in Washington. She was told that the last anyone saw of me was getting into a taxi en route to the airport. Although she was worried and in a state of disbelief, she kept reassuring everyone who called that I was a survivor.

So, while I was relaxing on the Metroliner, my wife, family and friends had some agonizing hours. They thought I might be dead.

It was difficult for most people to believe that the Bell System was dead. The culture shock was dramatic and lasted a long time. It was hard for people to accept that this American institution was no longer what it had been for almost a century. Many asked: If it's not broken, why fix it? Among the employees, there was a general feeling of disbelief. They had been indoctrinated from the day of hire that there are two kinds of people in America, those who work for the telephone company and everyone else. And, now, the family was being divorced. It couldn't happen, they thought, hoping that the government would put the Bell System back together again. [With the rash of mergers in 1996-1998, perhaps it will be put together again, almost.]

The period following the approval by Judge Greene of what became the Modified Final Judgment (MFJ), but before its actual implementation, was a time of great uncertainty. There was confusion among the public, not knowing whom they would call when they had problems with their phone service. CWA members faced tremendous stress because they didn't know how the breakup would ultimately affect them. Thousands of telephone workers realized they would go to bed one night and wake up the next morning working for a new employer even though they still had the same basic job. How secure would those jobs be now that the Bell monopoly was breaking up?

During that period, I went on the road and visited every Bell local in District One, giving a talk on how divestiture would affect CWA workers. I gave a slide presentation and answered their questions. Through the course of my road show, I personally spoke to about 50,000 members.

The members were very interested in what I had to say, particularly since there was no information forthcoming from the company. There were massive turnouts at meetings. Clearly people were concerned about their jobs, their futures and their families.

My presentation took almost two hours. During that time, nobody ever left, even to go to the bathroom. At one meeting in Queens, they forgot to shut down the bar. I proceeded with great trepidation, but nobody got up to get a drink. That showed the intense interest and fears the members had. After I finished my talk and opened the floor to questions, there would often be dead silence. So I would say, "Now, don't tell me you're not concerned about whether to sell or hold your stock." That usually got people going. And once the first question was asked, they had many more.

The question most members asked was, if they had the option to choose which company to work for, what should they do?

And here I unfortunately gave them some bad information which spoke to the uncertainty of the times.

The MFJ said employees would follow their work. But many workers had the option of either staying with the Regional Bell Operating Company (RBOC) or going to AT&T. I told the members that if they were within a few years of retirement to stay with the operating company. I didn't think anything very exciting was going to happen as they became dial-tone companies. But I believed AT&T was the place for the future. They had long distance, the best research lab in the world, the best factories, and a world class workforce. They would be the only company in the world that could offer one stop shopping, telephone service, manufacturing and research and development. I told workers with their careers ahead of them to go with AT&T.

I was wrong. AT&T was not as solid as I had thought, and went through great turmoil following divestiture.

Part of AT&T's troubles was the direct result of the Reagan Administration's short sighted economic and trade policies which had devastating effects on AT&T and our members.

One of the forces driving the antitrust suit was the objective to stop AT&T from buying all of its equipment from its own subsidiary, Western Electric. The Reagan Administration's theory was that if the company was forced to purchase equipment on the open market, prices would be lower and thus telephone rates to the subscriber would come down.

At that time, the twenty-two Bell Operating Companies purchased all of their switches, telephone gear and even paper clips from Western Electric. Notwithstanding the fact that the Reagan Administration knew that the rest of the world—in particular Germany, Japan and Canada—were waiting for an opening of the U.S.

telecommunications market, there were no U.S. negotiations with any of these countries for a reciprocal trade agreement.

Thus, virtually overnight, AT&T lost about 50 percent of its domestic market share. Market share and jobs are related. As market share drops, so do jobs. When AT&T lost market share, we lost jobs. Since overseas markets were closed to them, they had no alternative but to close factories and layoff thousands of workers. It was during this time that the United States also went from a favorable trade balance to a trade deficit in telecommunications for the first time in history.

Throughout the succeeding years, the playing field was anything but level. AT&T, as the dominant carrier, was under heavy regulation while its competitors were given advantages in the name of competition.

One stop shopping, which we all believed was a strength, turned out to be a detriment to AT&T's growth. The domestic market was more and more difficult for AT&T to sell to because the RBOCs were finding themselves more and more in direct competition with AT&T in both the marketplace and the halls of Congress. Thus, the RBOCs were shying away from buying equipment from AT&T.

The spinoff of AT&T's manufacturing operation into Lucent Technologies in 1996 resulted in renewed business with the Bells and augurs a bright future for Lucent and our members who work there. Hopefully, it also put AT&T in a better position to focus on its strengths and finally provide the kind of stability we had anticipated at the time of divestiture.

Divestiture radically changed our collective bargaining relationships. Prior to the effective date of divestiture on January 1, 1984, we had our last national bargaining with AT&T in the summer of 1983. To say that it was different from previous negotiations would be an understatement.

It is rare, indeed, to see a lack of discipline on a management bargaining committee, particularly in the old Bell System. The members of the management team had all been assigned to the companies for which they would be working come the following January. It seemed that each of them had their own agenda, something the new company would want or not want once it became independent of AT&T.

The 1983-86 contract was settled only after a twenty-two day strike involving 500,000 workers. This was the first nationwide Bell System strike in twelve years—and the last.

The strike was essentially over health care. For two months, the CWA members on the joint health care subcommittee pressed the company not to propose their new health plan. The company was told that their proposal of health care cost shifting would not be accepted. Three days before the expiration of the contract, the company formally put their health care proposal across the bargaining table. I didn't believe then, and don't today, that the companies were united behind this position. Unfortunately, it took twenty-two days on the picket line to persuade them not to tamper with our health care.

The 1983 contract settlement put us in a better position to face the coming breakup. But it was also the end of an era, the last time CWA bargained nationally for the entire Bell System. The period of national bargaining, 1974-1983, were the "golden years" for our members. Our contracts were among the best for union workers, right up with the best in the auto and steel industries. Major gains were made in working conditions, health and welfare benefits. In 1980, we negotiated a 30-and-out (regardless of age) non-contributory pension plan.

When I became CWA president in 1985, the first contract negotiations in the post divestiture era were less than a year away. We had to plan for a future for which there were no blueprints. It would be a whole new way of bargaining – for us and the companies.

This was also the period following President Reagan's busting of the air traffic controllers union (PATCO), when much of organized labor was losing ground in concessionary bargaining, two-tier wage systems and givebacks.

During this time, President Watts commissioned a poll of our members and the results were devastating. The vast majority of our members were ready to accept a status quo contract. They wanted us to go in, get a contract, and not have any trouble. Almost 15 percent were willing to give some concessions.

Armed with this knowledge, between August 1985 and February 1986, I visited our members in some 30 states. I talked about their employers, the wealth the companies had accumulated, the workers' value to the company and emphasized that not a single telephone company could make a case against a good contract, much less for any concessions. Everything I said was meant to bolster morale, create optimism for the future and be ready to fight for what was rightfully theirs. Our effort was largely successful. CWA members approached the 1986 bargaining with strengthened resolve.

AT&T was the first one out of the box. The company came in with several concessionary demands involving health care and other items. Negotiations were complicated even further since after divestiture, people followed their work and we had to merge many contracts into the AT&T agreement. Of course, the union wanted to level up all conditions, while the company wanted to push everything down. This was a tough negotiation, taking a great deal of time and compromise. It was also the first time we would negotiate wages for our members in the new AT&T. During the days of national bargaining, in order to avoid "mismatches" of wages for the same job in the same city, our members in AT&T received the same wage increases as did their counterparts in the operating companies. This was now gone, and each bargaining unit had to be negotiated separately.

During that collective bargaining session Bob Allen became the AT&T CEO. Shortly thereafter, AT&T Vice President Ray Williams, a long-time friend of mine in whom I had the utmost trust, told me that his orders were to get the cost-of-living adjustment clause out of the contract. The driving force behind that demand was the fact that all of AT&T's competitors were non-union and did not give their employees cost-of-living adjustments. Thus, AT&T's competitors could predict their costs in the years ahead. Not knowing where the cost of living would go, AT&T would be at a disadvantage, he said.

I told him that their demand was unacceptable and would result in a strike. This union could never give up the cost-of-living adjustment guarantee without a fight. And it did trigger a strike that lasted 26 days.

On the second day of the strike, I wrote to every member. I gave them a telephone number to call and leave any message they cared to and promised that within 48 hours someone would get back to them. We received and returned some 6,000 calls. The calls told us a great deal about our members and their concerns during this difficult time.

We learned that a vast number of our members were in financial distress even before they missed their first paycheck. This was largely among women who were single heads of households. Because of the nature of the industry, at least at the time, AT&T employed a larger percentage of female single heads of households than virtually any other company. We needed to learn how to better serve these members.

The call I remember most vividly was from a member in the Midwest. He said that his wife had called earlier to tell us that their electricity had been turned off for lack of payment. He said that while it was true, he was sure other CWA workers were in worse shape—we should take care of them and he would take care of his own problem. This spirit of concern for others prevails throughout our union and is a great source of strength to me.

A major step to help all CWA members who are forced to strike was taken by 1990 CWA Convention delegates when they voted to establish the Members Relief Fund (MRF). It is funded by member contributions of 0.15 percent of wages. The MRF is at $170 million and growing. It now grants each striker $200 a week. However, I can see the day when the MRF could replace fifty percent of each striker's after tax pay, maybe more.

During the 1986 strike, I had to make my first real hard decision as president. After twenty three days, about 25 percent of the strikers had gone back to work. We knew that if the strike did not end that weekend, by the following Monday many more would drift back. So we worked hard to get everything settled, knowing we would have to give up on the cost-of-living adjustment.

We succeeded in negotiating an agreement, but our rank and file bargaining committee refused to go along. They voted to continue the strike by a majority of one. After learning the various concerns that caused the local committee to vote to continue the strike, I managed to get the company to agree to resolve some of them. We advised them of the company's concessions. Later that day, they held another vote. Again by one vote, they voted not to end the strike, but this time one who had voted yes had switched to no. I knew that the time had come for me to exercise my responsibility as president.

Under the CWA Constitution, the Executive Board has the authority to terminate strikes. Our Board members had been kept current on the situation and during a telephone conference call, I proposed that the Board terminate the strike. With, I believe, two dissenters, we voted to end the strike.

Negotiations then proceeded with all of the newly spun-off Regional Bell Operating Companies (RBOCs). All of them were contentious in different ways and several took short strikes to get the final agreements.

During the 1986 bargaining we learned just how independent our different employers now were. No longer could we kick New York Telephone in the shins and make AT&T feel it. Nor could we pressure one operating company to win a point at another. Now the telephone companies were genuinely independent of each other and they couldn't care less how we treated their former colleagues.

We learned several other lessons. For the first time, we were impacted by the status of regional economies. This was not a concern when we bargained for the entire nation. The difference hit me when, after all the contracts were in, I went to a local union membership meeting in Davenport, Iowa. When I arrived, the meeting was underway and a young member was speaking. He said, "I appreciate what the union achieved in this contract; that if my job is eliminated here, I can go to another location in another state with no loss of pay. But," he said, "I need advice. Should I quit my job and keep my house? Or should I abandon my house and go to my new job location?"

The farming industry in the Midwest had gone belly up and our members there were being directly affected. We also felt the effects of regional downturns at Southwestern Bell where the oil industry was in a recession, particularly in Texas and Oklahoma. And the state of Texas had laid off 10,000 public employees during the time we were bargaining.

When most CWA members were employees of the Bell System, we were impervious to these external factors. Now we had entered the real world. The era of the Bell monopoly was really over in the sense that the philosophy they ingrained in new employees was no longer applicable. The old Bell maxim that there were two kinds of people in the United States—Bell System employees and everyone else—now had a very different meaning for us.

Clearly, CWA members were now part of the everyone else. We needed to change our ways to face the new realities of a post-divestiture world. And it wasn't going to be easy.

CHAPTER FIVE

Mobilization

I learned about the power of collective action when I worked for Mackay Radio. Mackay, later American Cable & Radio Corporation, was a subsidiary of ITT, an enormous company in a highly competitive industry. The employer could lose an important client because some competitor delivered a stock quote in London seconds before we did. The pressures on us were tremendous.

Following our failed strike in 1948, the company began speeding up the circuits, having the foreign countries send the telegraph messages faster so that we would have to transcribe them faster. The boss put my friend and coworker Mike Mignon on the fastest radio circuit—the one from San Juan, Puerto Rico, which served as the hub for all Latin America.

Mike was one of the founders of the ACA, and a leader of the workplace resistance, which is why the company put him on the hot seat. As the telegraph tape kept coming in at the faster speed, Mike worked at the normal speed. He quickly fell behind. I was told to relieve Mike, and after ten minutes at the circuit, the telegraph messages had piled up even more. I was finally relieved and the supervisors found someone to take my place. But he, too, resisted. This went on until the supervisors finally got the point and slowed things down to the normal speed.

It was a lesson that I never forgot. Today we call that type of action mobilization. But it is basically the same thing: resistance and solidarity in the workplace.

The first widespread use of membership mobilization in CWA came from our public sector members in 1986. Public workers in New Jersey are barred from striking even though they may bargain on economic and non-economic issues with their employer.

Not having the right to strike makes bargaining very difficult for the workers. The employer has very little incentive to come to a deal with the union. Delay in this case plays to the hands of the boss because the members get very frustrated if the union can't negotiate a contract before it expires.

We were forced to search for an alternative to the strike to effectively represent our public sector members. Thus, membership mobilization was born.

In 1981 CWA won representation elections for 36,000 New Jersey State workers in administrative, clerical, professional and supervisory positions. This followed a six-year organizing campaign that escalated after a four-day walkout in the summer of 1979. The walkout of more than 30,000 state workers was unprecedented, but threatened with the loss of their jobs, the workers returned without the pay increase they had sought.

CWA strategy in the organizing drive relied on building a grass-roots union, based on member mobilization around contract and legislative issues. Much of this was driven by the New Jersey State Supreme Court decision in 1979 that the state could fire workers who strike. More importantly, it was based on using the power that state workers could employ while staying on the job, particularly in a prolonged effort.

Faced with extremely tough negotiations in 1986, CWA-represented state workers formed a "Committee of 1,000" mobilizers from all eight CWA locals. Committee members received a special pin in exchange for pledging to be responsible for at least ten workers at their work sites throughout the contract effort. They also agreed to attend two statewide training sessions held on Saturdays at a state college.

A variety of mobilization activities followed, including tens of thousands of postcards sent to the governor on key issues, and huge lobbying days where CWA shirts were worn by 1,000 or more members meeting with their legislators. Similar meetings were also held in most of the twenty legislative districts.

In the end, because of significant member involvement, a good contract was negotiated. Within two years, mobilization became a national program in CWA.

Before divestiture, while most contracts in the telephone industry were settled peacefully, there had been some long and bitter disputes, including CWA's seventy-two-day strike against Southern Bell, IBEW's long strike against Illinois Bell during the Democratic Convention in 1968, and the granddaddy of them all, CWA's 218-day strike against New York Telephone. The independent companies were hit as well. When I was vice president of District One we had six-month strikes against Rochester Telephone (NY) and United Telephone (now Sprint) in New Jersey.

After divestiture, automatic strikes at contract expiration were no longer effective. While we would never give up our right to strike, we needed to be less predictable and come up with alternatives to strikes. Coupled with this was the need to involve our membership in all aspects of life in the union. This had not been as important when we dealt primarily with the Bell System. But now membership involvement would be critical in future dealings with the divested industry.

Following the end of the 1986 round of bargaining and as I began to attend district, regional and local meetings, I talked about how predictable we have been for almost fifty years, going on strike the minute our contracts expired. I would continue by saying that while we would always retain our right to strike, we needed to find alternatives to strikes. At every single meeting, that remark was greeted with spontaneous applause. I knew our members were ready for change.

Thus, we began the process where mobilization became an integrated form of membership communication and action throughout our entire union. Mobilization is organization within the union - the ongoing process of getting all of our members more involved in the union, in the workplace and in their communities.

With mobilization, we now have new methods of exerting union power. By using a mobilization campaign instead of a strike to pressure employers and show solidarity, we are able to make our point while still remaining flexible. We keep our members on the job, where they continue to get their paychecks and benefits and can actually have a greater impact on the company's operation than standing outside on the picket line. This is particularly true when a strike does not shut down the employer.

Disagreements with management over wages, health care and other primary worker concerns have always been strike issues with us, and will always remain strike issues. CWA and its members will always be ready to go out on the picket line. This was shown by the seventeen-week strike against NYNEX in 1989 to protect health care. But we have also learned that the strike should not be the only weapon in our arsenal. Sometimes we can gain more by not striking, or striking by other means. Instead of walking out of the workplace, our members can remain on the job, using innovative mobilization tactics to demonstrate solidarity, voice their concerns, and pressure management to change its position.

We need to mobilize in order to maintain and increase union power, and we need to always seek new and creative ways to demonstrate that power. Unless we are effectively mobilized, our employers will be able to communicate better with our members than we do. Education is absolutely the key. There has to be constant two-way communication between leadership and members.

Since the beginning of its history, CWA has always been an active union. We were organized on the basis of member involvement. There was a great deal of one-on-one contact between union leaders and the members, information flowed both ways, and political power often came from the bottom up, rather than the top down.

Over the years, we lost that sense of participation. Members became passive, relying on leadership at both local and national levels to solve their problems for them. Unless it was contract time, or the union was proposing a dues increase, there often was not much union activity. This was not peculiar to CWA, but happened throughout the labor movement.

I noticed this passivity at Labor Day parades, when there were more spectators on the sidewalks than union members in the march line. When I was elected president in 1985, I set as my objective to turn that around, stating the policy of our union was one of activism and inclusion. We wanted and needed every member to be an active participant, so that we would all have a stronger union. Our members had to take a more active role in the union, the workplace and the community. Mobilization was at the core of this revitalizing change.

But change comes slowly in a large organization. When I first raised the idea of an electronic picket line, or a boycott of AT&T, during 1986 bargaining as an alternative to a strike, the idea floated about as high as a lead balloon with the local leadership. After the 1986 strike, we knew an intense education effort among the local union leadership was necessary if the membership was ever going to buy into the strategy of mobilization.

We began to develop the electronic picket line immediately after the 1986 walkout ended. And, as I explained it, I could tell that the vast majority thought I had lost my mind.

I would point out how unions put up picket lines outside of establishments to keep customers from going in and thus, hurt the employer's pocketbook. Well, when we put picket lines up around a telephone building, we didn't cause our employer to lose a single dime. So, we had to cause AT&T customers, for example, to use another carrier.

The old Bell System indoctrination came to the forefront. Those with Bell Heads thought it was terrible for their president to tell customers not to use our company. It took an awful lot of repeated discussion before people began to understand that the electronic picket line had virtually the same effect as a strike with less risk and pain to our members. Here was a way to cut the employer's revenue while remaining on the job.

Our arguments finally prevailed and during the 1989, 1992 and 1995 AT&T bargaining sessions, we used the electronic picket line. We collected tens of thousands of authorization forms from AT&T customers that would permit us to transfer them from AT&T as their primary long distance carrier to another long distance company. We also printed several million small stickers, the size that could be placed on a telephone, that gave the caller the five digits necessary to bypass AT&T.

Jim Irvine, our Vice President with AT&T responsibilities, and I held a news conference and showed the gathered press what we had accumulated and what we were prepared to do. We got the same reaction from the company management as we did three years earlier from our members. How could this union, AT&T asked, even think of giving our business to a competitor?

This pressure tactic, in concert with all of our other mobilization tactics, visibly helped our bargainers achieve a satisfactory contract with AT&T in 1989. Indeed, prior to the contract signing, we had a mailing prepared with the assistance of the AFL-CIO to nearly one million union members with long distance carrier switch authorization cards ready to drop into the post office. It was one of the best mobilization campaigns we had ever planned that, fortunately, was not necessary to use.

The 1989 bargaining session was also the first time CWA and IBEW bargained jointly with AT&T. On the day preceding the 1986 strike, AT&T Vice President Ray Williams told me there would be one more offer coming prior to the midnight expiration. That offer never came because, unbeknown to us, IBEW signed an agreement for the workers they represented. We, nevertheless, struck. IBEW members worked.

Later that year, Williams and I were invited to speak on several occasions about these negotiations. In January 1987, we were at Pace University in New York City. For the first time, he said, in these negotiations, the company acted like any other company in the competitive marketplace and took advantage of the split between our two unions.

All I could say was, "It will never happen again."

I went back to Washington. That night there was a dinner honoring the retirement of IBEW President Charles Pillard. I sought out the incoming president, a New Yorker named Jack Barry. I told Jack what had happened that day and said that we could not let it happen again and that he and I, as new presidents, have an opportunity to forge a partnership between CWA and IBEW. He fully concurred.

This led to me assigning CWA Executive Vice President John Carroll and Jack assigning his assistant, Tom Hickman, to work with Ray Williams to forge a mechanism for joint bargaining. John Carroll did a magnificent job, but retired prior to the 1989 round and died shortly thereafter. He was a pioneer and great trade unionist.

For most of the time between 1986 and 1989, every time he saw me, Ray Williams said: "There will be health care cost shifting in '89." My response was always: "Like hell!"

Faced with the prospect of either a strike or the transfer of tens of thousands of their customers to other companies, or both together, AT&T withdrew their health care cost shifting demand three days before the expiration of the contract and was willing to maintain the status quo. However, Tom Hickman and I refused to accept that.

In 1988, AT&T spent $1.5 billion on health care. It would rise to $1.9 billion in 1991, and $2.5 billion in 1994, while providing the same level of benefits. We knew that maintaining the status quo would just put off the day of reckoning and we would rather deal with rising health care costs now. We actually dragged the management, kicking and screaming, into agreeing to develop the first nationwide managed health care program, jointly built and operated. This, as well as what was done with all of the companies except NYNEX (now part of Bell Atlantic), tended to keep health care increases to the low single digits, while delivering quality care to our members.

The support of our members played an enormous role, not only to deliver the AT&T contract, but to impose the parameters of that settlement throughout the former Bell System. It also had an impact on what we were able to do with the former "independent" companies, GTE and Sprint.

While we continue to adapt to the changes wrought by divestiture, we had learned many lessons from the process by then. One thing that divestiture ultimately taught us was the power of our members. Prior to 1984, the union's power largely resided at the top. We didn't really need our members to be active. Divestiture and the competitive marketplace changed all that. We realized we needed a more active, more militant and more educated membership. It was our role as union leaders to help members empower themselves.

In our 1992 bargaining with AT&T and Bell Atlantic, we worked beyond contract expiration in both companies. Meanwhile, we conducted mobilization activities and kept the companies off-balance, because they never knew if or when we might walk off the job. One way in which this campaign proved a tremendous distraction was that it kept Bell Atlantic busy making and changing hotel and travel arrangements for management personnel they thought might be needed in the event of a strike.

During the campaign against AT&T, we conducted a nationwide "Stand Up To AT&T" action. On April 15, 1992, at the exact same time all across the country, CWA members at AT&T workplaces, most of them wearing red, stood up on the job for five minutes.

By the 1992 contract talks, our members also were ready to incorporate a nationwide electronic picket line campaign as a key element of our bargaining strategy. We knew that about seventy percent of all union families, more than nine million subscribers, had AT&T as their primary carrier. In addition, all 75 national unions affiliated to the AFL-CIO, the 50,000 local unions, the AFL-CIO and all its offices across the country and union friendly vendors, all chose AT&T because it was the only union long distance carrier. This business was estimated to be about $300 million annually.

Also, what if we could get a large number of AT&T employees to switch carriers? The public relations value alone would be enough to severely embarrass the company. Just imagine how their competitors would ridicule AT&T in the marketplace when their own employees switched carriers.

We printed and distributed self-adhesive stickers advertising long distance alternatives to AT&T. Those stickers were on public pay phones for months afterwards, even though the issues that provoked the response had long since been resolved.

To help bolster the electronic picket line, we used the Strategic Approaches Committee at the AFL-CIO. We generated photo-ready copy that could be included in every AFL-CIO union newspaper telling workers how to bypass AT&T. All union editors were alerted. We also had access to those 13 million union homes. Most importantly, AT&T management knew we had all of this in place.

We also found ways to exert boycott-type pressures on the regional operating companies. Because there were no competing operating companies for subscribers to switch their business to, we came up with other effective tactics. Realizing that services like caller ID and call waiting are very profitable lines for the regionals, we coordinated a boycott at both the national and local level, where CWA staff and local officers and members asked all its vendors, colleagues, and local businesses to sign authorization forms to drop these services in solidarity with CWA workers. This campaign added up to a potential loss of millions of dollars in revenue for the regional operating companies and added another weapon we could use against them.

One of the best things about mobilization is that employers hate it. In fact, our employers have told us that developing a contingency plan to counteract our mobilization campaigns cost them millions of dollars.

As a result of our mobilization strategies, some companies have come to us and requested to bargain early and even agreed not to propose any demands upon us. They were making concessions before bargaining even began in order to avoid having to deal with our mobilization. Just as the threat of a strike can be as effective as striking itself, we have made the threat of mobilization as effective as a full-fledged campaign. And, we continue to improve and perfect the process.

Mobilization took a new twist in 1995. NYNEX and CWA reached an early three year agreement in 1994. Bell Atlantic negotiated early with two IBEW locals that represent a small minority of the company's employees. They reached agreement on a substandard five-year contract.

Bell Atlantic CEO Ray Smith bragged to his colleagues that he showed them they didn't have to follow the CWA-NYNEX contract terms. Negotiations were confrontational from the first day. Management negotiators told us if there was no agreement by expiration day, they would not extend the contract, payroll deductions for union dues would end as would the provision for arbitration of grievances.

And, it happened.

To nullify the company's objective of financially strangling our local unions by cutting off their dues, I notified each local that the national union would advance them their usual dues rebate, to be paid back after we got a contract. This enabled locals to devote their full energies to mobilization.

What brought the company to the bargaining table and to bargain in good faith was a $7 million media campaign, mostly television advertisements, in key Bell Atlantic cities. The message was designed to embarrass the company, and was used to augment other mobilization activities. And it worked spectacularly.

The ads featured an incompetent contracted-out telephone repairman and installer by the name of "Larry the Contractor." Larry was a telephone customer's worst nightmare: unkempt, disheveled and totally inept. When he came into your home, he would eat out of your refrigerator and break your favorite possessions. And when Larry left, your telephone was always in worse shape then before he arrived.

Our point, of course, was the deterioration of service quality in Bell Atlantic because of the company's widespread use of contractors. I can't tell you how many calls I received saying, "Larry was in my house yesterday." We had really struck a responsive chord with Bell Atlantic's customers.

The company went bonkers over the favorable public response to our ads. The ads and the earned media attention that we received helped bring the two Bell Atlantic vice chairmen to the bargaining table with me and our two vice presidents, Pete Catucci and Vince Maisano. That meeting was a first, and after five months, the company agreed to our terms.

What is noteworthy is that during the ordeal, two other Bell CEOs told me if we let Smith off cheaper, "Look out for me next time." I let Bell Atlantic know they had said that. So, the industry raised the stakes and that was fine with me.

Also noteworthy from that experience was that the $7 million spent on the media campaign is one week's strike benefits.

Mobilization in CWA today takes many forms. From lunchtime rallies, to mass sick-ins and information picketing at the homes of managers. Membership education is a vital component. By education I mean two-way communication between the mobilized members and their leaders. The leaders must educate the members: If members do not understand the issues at stake and how they might affect them, they will be less willing to get involved. And the members must educate the leaders by making them constantly aware of the workers' concerns, how the mobilization campaign is progressing, what impact it has, which tactics are effective and which are not.

Ideally, education is achieved through one-on-one contact, which requires a strong organization within the union. The CWA mobilization plan calls for each local to have a local mobilization coordinator, each work site location to have a building mobilization coordinator, and for every group of ten to twenty workers there should be a workgroup coordinator. (In work sites where there are only a handful of workers, there is no need for a group coordinator.) Members without formal titles also play important roles in mobilization campaigns, engaged in discussions about the issues and participating in collective actions.

Mobilization also occurs outside the workplace, in the community and the political arena, generating grassroots action.

Mobilization gets our members out into the community, creating coalitions with other workers and groups in order to advance a broad common agenda—the rights and aspirations of working families. Sometimes these coalitions deal with issues that directly affect our members. During one bargaining session, AT&T made proposals that we felt were antiwoman, so CWA built a coalition with the National Organization for Women, the National Women's Political Caucus, Coalition for Labor Union Women and other women's groups which rallied to our cause.

Other times our community mobilization efforts are on behalf of wider goals of social and economic justice, often concerning issues that might not directly affect CWA members. Nowhere has such broad-based community mobilization been more successful than in the case of Jobs with Justice, a grassroots based coalition that seeks to promote workers rights and economic fairness.

Jobs with Justice has helped workers win organizing and bargaining fights that might otherwise have ended in defeat. It has rallied tens of thousands of people in nationally coordinated strike support activity. Jobs with Justice also mobilizes in support of a wide variety of issues that concern, not only union workers, but working families, the poor and minorities.

Founded by CWA Organizing Director and Assistant to the President Larry Cohen in 1987, Jobs with Justice serves as a model for grassroots organizing and community action. It has already built powerful and active coalitions among trade unionists, students, senior citizens, consumer groups, environmental activists, religious leaders and other concerned citizens. Because of its strong coalitions and bottom-up organization, Jobs with Justice can generate a great deal of "street heat" on short notice.

One way to make certain that Jobs with Justice has both the

organization and the commitment to be effective is the "pledge to be there." Based on the old Wobbly principle that an injury to one is an injury to all, Jobs with Justice asks its coalition members to pledge that they will personally participate in at least five actions a year for other coalition members. The pledge gets union members out and involved in support of broader issues and it also gets other coalition members out and involved in support of union issues. The pledge also reinforces the idea that solidarity goes both ways, anyone who receives help must give something back to others.

Our first major Jobs with Justice campaign was launched in 1987 in conjunction with CWA's convention in Miami. Fifteen thousand activists attended a rally in support of the Machinists strike against Eastern Airlines. I had also invited all of the Democratic Presidential candidates to address our convention. At the time, the National Association of Broadcast Employees and Technicians (NABET) was on strike against the local NBC affiliate so I barred the NBC camera crew and reporter from the convention center.

The following day I was served with an order from a U. S. District Court judge requiring me to give the NBC crew entry. A contingent of U. S. Marshals was dispatched to the convention to hustle me off to jail if I did not let the NBC scabs in.

I got advice from our legal counsel, but did not follow it. Instead, I barred all media coverage from the auditorium. I was later cited for contempt for not following the court's instructions.

We got through the convention and the issue would have gone away, but I didn't let it. Instead, I instructed our general counsel to file an appeal. He said: "Are you serious? The appeal will be in Atlanta, which is hardly a union environment."

But occasionally justice triumphs. An Appeals Court upheld our right to bar scabs from covering our convention under the circumstances that existed. NABET later awarded me a lifetime membership card. Little did we realize that this initial Jobs with Justice campaign would end with NABET turning to CWA as a merger partner several years later.

Mobilization shows that when we work together, we build a union that is much stronger and more effective. And it strengthens the union in its role in collective bargaining. Mobilization is particularly critical in winning first contracts. While there is a euphoric feeling when the votes are counted in an organizing campaign and the union is victorious, reality sets in the next morning. The fledgling union now has to negotiate a first contract.

Statistics in recent years show that as many as half of labor's organizing victories never end up with a contract because the employer's resistance continues. Not only must the activity that led to the victory continue among the employees at the organized enterprise, but other CWA members and, indeed, the AFL-CIO state and local organizations must be involved in mobilization activities. Effective mobilization campaigns result in substantially more first contracts.

Mobilization and alternative strike strategies are our way of empowering members to work together with their local and national leaders to achieve a satisfactory result for all of our members. We learned that each of us needs to be ready to be there for someone else's problem or fight. Because that same fight could be ours, next year. That is what we mean to be a part of the CWA family. Mobilization is even more important today, as corporate power becomes increasingly excessive.

CHAPTER SIX

The Struggle

Labor historian Jack Barbash once pointed out that "No political democracy has offered a more hostile environment to unionism than the United States. And this hostility has imparted to American unionism a character and temper unlike those of any other labor movement."

From my experience, he is correct. The development of the U.S. labor movement is quite different from those of other industrialized democracies, and CWA is shaped by the American labor tradition. Many people, even union members, fail to grasp the unique role of American labor unions in our economic system.

In many societies, unions were (and in some cases still are) engaged in a class struggle on behalf of workers against the holders of capital. Unions express the voice of the disenfranchised against those with power. That is why unions in most industrialized democracies became an integral part of a single political party. U.S. labor, of course, has evolved as a force separate and independent from the two political parties. But that's not the only difference between American unions and the labor movements in other countries.

The class struggle here in America is much different than in the rest of the world. It can best be described as a clash between the haves and the will-haves. The average American worker has an optimistic view of the future. Most workers, particularly union workers, expect to receive their fair share from the fruits of their labor. They don't necessarily oppose the capitalist system. They just want to get more out of it for both material reasons and personal satisfaction. The union is there to help them achieve their goals and to act as their voice on the job, in the community and in the political process.

Many within our ranks have suggested that labor could better represent our existing members and recruit new ones by starting our own labor party. This is an idea that has floated around since almost the beginning of the modern labor movement and I disagree with it. Joe Beirne, in expressing his opposition to a labor party, once correctly pointed out that the Democrats and Republicans are "parties of opportunity rather than parties of ideology."

He continued: "Third party movements throughout our history have discovered, to the dismay of their more ardent ideologues, that when they have developed sufficient support for a concept, that concept is apt to be taken over, lock, stock and barrel by one or both political parties."

I'm reminded of the Progressives, one of the most successful American third party movements, which elected many office-holders earlier in this century. Theodore Roosevelt and Robert La Follete were among the best-known Progressives. But their platform was largely co-opted by the Democrats and they have since faded into memory. The last successful third party movement was the Republican Party, which came to power directly as a result of the cataclysmic events leading up to the Civil War.

Another compelling argument against a labor party is the nature of American politics. Our electoral system is winner take all. All of the other industrialized nations have adopted parliamentary and other power-sharing political systems where a party can win some of the vote and gain seats of power. American politics, more so than other nations, is fueled by enormous sums of money. With elections taking place every year at the local, state and federal level, it is simply naive and unrealistic to expect big donors to fund a fledgling political party that would take years, if ever, to gain some measure of power for the investment.

A labor party is not in America's future. Union leaders best serve our members by remaining an independent force and being active in all levels of American life.

CWA's relationship (and that of all labor) with our employers and the business community operates within this economic, social and political context. But while unions have often been accused of fueling class warfare between workers and their employers, quite the opposite is true. The business community historically fights unions. They see unions as a threat to their property rights and an obstacle to their doing as they please.

The prevailing view of labor relations in the mid-19th century was that the employer paid wages for services rendered and had no other responsibilities to the workers. Organized labor threatened that power arrangement. Early labor history in the United States was marked by heroic struggle, often against long odds, with episodes of violence and the involvement of hundreds of thousands of workers. It is compelling history. Of all the national craft unions formed in the 19th century, the Typographical Union, founded in 1852, is the oldest. CWA is proud to be the home of the ITU and a keeper of this heritage and a living link to labor's past.

Throughout the 19th and early 20th centuries, the business community engaged in a belligerent campaign to crush labor.

But this early struggle established a beachhead for unionism in America. The events of the Great Depression led to the passage of the National Labor Relations Act, which led to the birth of CWA. Enacted by a sympathetic Congress at the initiation of the Roosevelt Administration, the NLRA was hailed as labor's Magna Charta and unions grew dramatically in the period that followed. Unionism appeared to be on its way as an accepted facet of American life.

But this success was short lived.

America's split personality toward labor revealed itself when workers joined unions by the millions and workers' power became a force to be reckoned with by business. Congress adopted two crippling amendments to the National Labor Relations Act, the 1947 Taft-Hartley Act and the Landrum-Griffith Act in 1954. Although not targeted specifically to put unions out of business, these highly restrictive laws were clearly designed to limit union growth and power.

Where unions were entrenched or couldn't be replaced, particularly in rails, mining and the mass industries, an unspoken social compact developed as America enjoyed the economic riches of our victory after World War II. The social compact was an understanding that unions would be left to negotiate improved wages and benefits for their members while managers would be free to run their businesses in the manner they saw fit, to earn higher profits that would be shared in collective bargaining.

As the only industrial power to emerge unscathed after the war, American corporations dominated the world economy for nearly twenty five years. The social compact lasted as long as American hegemony over the world economy remained in place. During this period, union organizing went to sleep as a new form of "business unionism" crept into the labor movement. Too many local unions operated like businesses. Their leaders lost sight of our larger mission; instead they focussed too much attention on operational and financial matters. Union membership as a percentage of the workforce slowly declined. Unions continued to provide excellent representation to their members. However, the expected influx of new workers into the ranks of labor as a result of unions' unprecedented gains for their members simply did not materialize.

CWA was shaped by these times.

Like millions of other Americans after World War II, telephone workers were anxious to restore their standard of living after victory was won. In 1947, the NFTW led a nationwide strike that was a disaster. Out of the 1947 struggle, the NFTW evolved into CWA, a union with greater unity and strength. Because many of the NFTW unions remained independent and because of the competition with the IBEW, CWA retained a strong organizing focus. Organizing, in fact, was a specific responsibility of a CWA executive vice president.

CWA resisted business unionism. But with virtually all of our members employed by regulated monopolies, many of our locals fell into comfortable niches during the 1950s-1970s. Locals mostly took care of their members and protected their work from other unions. We lost much of the fire and spirit of our early days. As the telephone industry grew so did our union, without having to aggressively organize.

I became vice president of District One in 1969, just in time to lead the strike that would shake the telephone industry. Prior to 1974, negotiations in the Bell System were marked by pattern bargaining. The CWA Executive Board selected a bargaining unit to be the pattern setter. All other units marked time while the union's resources supported the pattern setting unit. Following agreement, the union was usually successful in imposing the settlement on all units.

In 1971, our 37,000-member New York Telephone Plant unit voted to reject the pattern. The issues were wages, pensions and other benefits. The strike lasted 218 days notwithstanding the presidents of our striking twenty-three locals being aware that New York Telephone did not have the authority to change pensions or other system-wide benefits. There were more people on strike at the end than the first day, as nonmembers joined the union and the strike.

The strike against New York Telephone posed an enormous dilemma for Joe Beirne. By rejecting the terms of the settlement, the presidents of the New York Telephone locals essentially rejected the pattern agreement already approved by the CWA Executive Board and accepted by every other unit. Joe knew that AT&T would make it clear to New York Telephone management that any substantial deviation of the pattern would destroy this kind of bargaining in the future, since it would render the pattern agreement meaningless.

Beirne instructed me to hold another meeting of the local presidents. I did, to no avail. When it became clear that the members supported the strike, Beirne authorized it, thus putting the entire union behind it.

On August 14, 1971, Beirne announced that the members had ratified the contracts across the Bell System, except for New York.

The next day, President Nixon announced a ninety-day wage, benefit and price freeze. Some observers believed that Joe somehow learned of the Nixon plan and so announced the ratification just under the wire.

President Beirne asked me to call a meeting of the striking locals. We met at Parkers restaurant near LaGuardia Airport. With him were Executive Vice President Gus Cramer and his (Beirne's) assistant, Dick Hackler. He told Dick to start the tape recorder. One local president stood up and said: "There will be no taping of this meeting."

Joe said: "I called this meeting. Anyone who wants to leave should do so. But, know that I will write to your members and let them know you didn't participate because you didn't want your words recorded."

No one left.

Joe explained the meaning of the freeze—that absolutely nothing would happen for 90 days. The presidents were not moved. There was a good deal of hostility against Joe.

Joe presided over similar meetings at Parkers every Friday for the next six months. As the weeks dragged on, we sensed the sentiment changing from hostility to a plea for his help. While the meetings were for the purpose of exchanging information and strategizing, Joe used them to win over the entire group.

Joe went all out to end the strike. Thanks to his leadership, the strike finally ended when Curtis Counts, chairman of the Federal Mediation and Conciliation Service, offered a proposal that we agreed to send out to ratification. The contract was in fact ratified and we won a major concession—the company agreed to union security for the first time. We won the agency shop, where nonmembers had to pay fees equal to union dues.

Joe Beirne was proud of the display of courage by so many for so long. At our board meeting prior to our 1972 convention, he asked our public relations director, Lee White, and me to meet with him. He said he wanted to start a seven months club, open only to those who were presidents of striking locals or members of the bargaining committee. The club would meet annually at the CWA convention until there was only one survivor. Ironically, Joe was the first to pass away. Today, there are only three of us left—Ed Dempsey, President of Local 1101, Fritz Clark, President of Local 1111, and myself.

White came up with a seven months club certificate. Joe held a ceremony at the next convention where each club member signed the certificate, with Joe standing by. Each got a photo of his signing. My certificate and the photo are displayed in my conference room with Joe's inscription: "To Morty Bahr—who had to sweat it through."

The 1971 strike is a proud part of CWA history. When 37,000 workers stand together for seven months—it is a victory. Such a display of solidarity led to an enormous change in Bell management thinking. In order to avoid another local company strike of this proportion, AT&T agreed to national bargaining in 1974.

We had achieved Joe Beirne's dream of nationwide bargaining in AT&T a few months before he died. The next twelve years were CWA's "Golden Era" as we negotiated some of the best collective bargaining agreements in the nation.

CWA members were largely insulated from the turmoil of the 1970s. Oil shortages, stagflation, inflation and the rise of global competition caused many unions to begin to lose members. As America's preeminent economic position was threatened, the social compact began to feel great strains.

In 1978, with a Democratic president in the White House and Democrats firmly controlling both houses of Congress, labor initiated a legislative effort to win modest changes in the National Labor Relations Act. President Carter promised to sign a labor law reform bill, but his Administration did little to help passage. With lukewarm White House support, we fell one vote short of breaking a filibuster in the Senate and sending the bill to President Carter for his signature. That one Senator was a Democrat who called himself a "friend of labor."

If the labor law reform bill had passed, much of the difficulties labor and working families experienced over the past two decades might have been avoided.

When Ronald Reagan became president, the old social compact was already crumbling. With a single act, he destroyed it. By firing the air traffic controllers and breaking their union in 1981, Reagan became the first president to ever destroy a union as an institution. He demolished all previous standards of behavior in labor-management relations and sent a message to corporate America that it was okay to bust unions.

The use of replacement workers to take the jobs of strikers was once repugnant to Americans. Management's right to use replacement workers was enshrined in a footnote in a 1938 U.S. Supreme Court decision. In the Mackay Radio and Telegraph Company case, the court said that workers could be permanently replaced for participating in a legal strike. Ten years after the ruling, I was working for Mackay and went on strike against the company. We lost that ninety-day strike. But no one was permanently replaced, nor did the company threaten it.

Mackay Radio, the very company that had won the right to replace strikers did not use it in 1948, nor in later strikes. Neither did ITT, Mackay's parent company. Up until 1981, the permanent replacement of striking workers was considered socially unacceptable behavior. It was viewed as too reprehensible, too divisive, far too likely to lead to violence on the picket line. Employers knew it was a major roadblock to the resolution of the strike. Reagan changed that unspoken promise by firing PATCO workers.

Now the use of replacement workers has become an accepted business practice. The breaking of labor law has escalated as businesses increasingly rely on antiunion consultants who know that the penalties for breaking those laws are either nonexistent or a cost-effective trade off. Outsourcing, downsizing and the use of cheap foreign labor have shown that company loyalty now only goes one way—workers must be loyal to their employers, but their employers don't have to be loyal to them. Corporate raiders use legal loopholes to break pension and health care promises to retirees. The needs of communities that had long nurtured successful companies were now ignored as plants and offices were shut down and moved out.

This was the world that CWA members were thrown into when the world's best telephone system was thoughtlessly dismantled to satisfy the ideological desires of the Reagan Administration. Like other union workers in the 1980s, telecommunications workers paid a terrible price, as an estimated 135,000 well-paying, high-skilled jobs, both management and union, disappeared. Fortunately, there was a safety net, the union contract.

Other unions which saw their basic industries deregulated had not fared well. Vicious labor-management conflict and devastating job losses occurred in industries such as airlines and trucking. In spite of a fifty year bargaining history in telecommunications, CWA suddenly faced more confrontational employers.

While tolerant of the union in its core business, telecommunications employers were fiercely determined to prevent any union expansion into the new businesses or joint ventures which they were now allowed to enter. Union workers were barred from transferring or seeking new jobs in the new subsidiaries. If laid off, they were forced to go off the employment rolls before applying for jobs in the nonunion ventures. A stone wall separated the union from its nonunion parts of the business; that wall was as impenetrable as the Iron Curtain that once fell over Eastern Europe.

Our worst suspicions were confirmed in 1991 after AT&T had purchased NCR (National Cash Register) for $6 billion. The purchase was made over our objections that it was a bad deal, and we were proven correct in 1996 when NCR—subsequently renamed AT&T Global Solutions—was spun off by AT&T after having lost billions.

Prior to the 1989 negotiations with AT&T, I agreed to meet with Bob Kavner, the head of AT&T's data network business. He laid out a rosy future of great growth and increased high-paying technical jobs. But he had a problem.

The technicians in this line of business had little seniority and were at risk of layoff. He needed a separate seniority universe.

Unions do not fool with seniority very easily. But we made the change in the interest of the business and our expectation to share a better future.

After the NCR purchase, these very technicians, some 1,200 workers, were laid off as AT&T transferred the work to the nonunion NCR. And Kavner didn't even have the guts to call me. I read his quote in the paper—as part of the takeover, he guaranteed NCR they would not suffer layoffs—so, our loyal AT&T workers got the axe.

The laid-off CWA members were allowed to transfer to the new NCR unit. But they first had to resign their AT&T jobs and go to work as new hires. They lost their pension credits, seniority and other accrued benefits. This was outrageous conduct.

We had started an organizing campaign in the new NCR unit in the face of hostile company opposition. An NCR employee gave us copies of an NCR management handout which outlined in detail the company's "union containment" strategy. The strategy called for management to "contain" the existing union in the company while maintaining a nonunion workforce in the growing parts of the business. Over a period of about fifteen years, the nonunion workers became the overwhelming majority of the workforce which greatly weakened the union at the bargaining table. It was only a matter of time before the union withered and NCR was virtually union-free by the time of its purchase by AT&T.

With top corporate support, the new NCR part of AT&T was committed to staying nonunion despite the five-decade relationship CWA had with the company. The battle lines were clearly drawn as the other telecommunications companies also followed AT&T's union containment strategy. The industry turned its union-containment face to us.

In AT&T, for example, the numbers of nonunion jobs kept increasing as the numbers of union-represented workers went down. In one campaign, we had bitter conflicts with management over our attempts to organize American Transtech, AT&T's telemarketing subsidiary in Jacksonville, Florida. In 1995, Transtech employed about 5,000 workers, of whom 1,200 were classified "management" and ineligible for union membership, 800 hundred were regular, full-time employees and 3,000 were temporary workers supplied by an employment agency.

CWA organizers discovered that most of the temps had been working forty hours a week there for three to five years. Though much of the Transtech work had previously been performed by CWA members in AT&T bargaining units throughout the country, hourly wages paid to Transtech workers ranged from roughly $6 for telemarketers to about $12 for the most complex service jobs—a fraction of the union-negotiated rates for the same jobs.

When Transtech workers attempted to organize a union, management waged an all-out assault. Transtech management defied a corporate policy to maintain a respectful relationship with the union and to remain neutral during organizing drives. Instead, the company ordered workers to attend meetings in which they were pressured to oppose the union. Transtech also required managers to distribute anti-union materials and to speak out against the union. Despite our best efforts and continued presence, Transtech remained nonunion until it was sold by AT&T in 1997.

This employer conduct is similar throughout American industry. CWA employers were no exception.

In 1991, we launched our "CWA: Wall-to-Wall" campaign to counter management's union containment strategy. "To those who might not have gotten the message, this means that CWA must be recognized by all of our employers for all employees in that company who can legally belong to the union. No exceptions. No excuses." I announced this in my convention address that year.

Throughout the 1990s we negotiated various forms of management neutrality in CWA organizing campaigns in telecommunications, card check recognition and standards of behavior. But our big breakthrough came in 1997 in SBC when management finally agreed to complete management noninterference in union organizing and card check recognition for all units in their business. The other major employers also accepted wall-to-wall contract provisions in 1998 bargaining and we will finish the job in our next round of bargaining, if not sooner.

When the new century opens, CWA will have achieved a significant organizing and collective bargaining goal that should serve as a springboard for us to move our wall-to-wall demands to every employer under contract to us. The enormity of this victory will become apparent in the years ahead.

The vast majority of CWA members recognize that a strong union is in their personal interest. They know that by expanding the numbers of union jobs in their company we also greatly expand their employment security. We have to make the same link in the minds of every worker we represent.

For the employer, a union workforce can provide a significant competitive edge as customer service, product innovation and quality take center stage in the marketplace. Corporate leaders who believe that their only responsibility is to the bottom line are sacrificing the long-term health of the company for short-term gain. Unfortunately, I have seen so much of that behavior from the CEOs with whom I have dealt over the years.

When former AT&T CEO Robert Allen announced 40,000 layoffs in 1996, the price of AT&T stock rose $2.125. But Wall Street was so captivated by downsizing, it got hoodwinked. The 40,000 layoffs were a fake. First, the number included several thousand layoffs that had previously been announced. Second, there was no way that AT&T could lay off that many workers without ultimately hurting the business. In reality, thousands of the layoffs never happened. AT&T actually finished the year with more workers on the payroll than when it began.

But the damage to employee morale was incalculable. And when Bob Allen's face appeared on the cover of *Newsweek* as one of the "Corporate Killers," his days as a leader in the industry were numbered. His service at AT&T will not be remembered with glory. In addition to the $7.5 billion NCR disaster, AT&T lost $150 million on Unitel, a Canadian long distance company that went bust, and had numerous other failures. The 1997 trivestiture of AT&T, in which its manufacturing business (which became Lucent Technologies) and NCR were spun off and a "new" AT&T emerged, was an admission of the company's failed business strategies since divestiture.

When a CEO fails, workers inevitably pay a price. While layoffs might create short-term profits, their net effect on the company and the economy as a whole are generally negative. What downsizing and reengineering have done is improve corporate profitability in the short term by reducing labor costs. But productivity growth on any great or lasting scale has yet to be seen. In the end, the squeeze on workers will only result in lower productivity, unless corporate America makes a commitment to create high-performance workplaces with high-skill, high-wage jobs.

I continue to watch with amazement, however, as top CEOs receive compensation packages that often reach grotesque proportions. Nowhere else in the world do CEOs make as much money as in the United States. I recently spoke to a corporate manager who told me that every time he gets his paycheck, he pinches himself. So, I asked him, "How much is enough?" He admitted that he was overpaid. Then he started making excuses that his company had to remain competitive. In order to keep the best people, they had to pay competitive rates. Other companies kept raising compensation levels, so his company had to keep up. I wondered why competition, in this case, made prices rise rather than fall.

The spread between CEO compensation and the average

worker's paycheck continues to widen. In 1965, the average CEO earned 44 times what the average factory worker took home. Today, they make 209 times more than the average factory worker. Since 1980, CEO compensation has increased by 500 percent. How much is enough?

Our members don't begrudge the boss for earning a good salary. But greed is way out of line when companies pay CEOs tens of millions of dollars while asking for more sacrifice from their workers. In our 1997 negotiations with ABC, which is owned by Disney, the NABET bargaining committee asked me to sit in with them. The vice president of the company started out by saying that the union just wanted too much—our demands were unrealistic.

I told him that I did not understand what he meant by too much, since just a few months prior they had given Michael Ovitz a $140 million severance package. Ovitz had only been with Disney for fourteen months and had done a lousy job, so he got $10 million for every month he screwed up.

Disney CEO Michael Eisner signed a $770 million contract in 1996 and in November, 1997 exercised stock options worth more than half a billion dollars. To put Eisner's stock options in perspective, his $565 million payout equals $1,074 a minute every day for the entire year. If $1,074 a minute is not too much for Eisner, why are CWA's contract proposals for its workers too much? Workers lose faith in the free market system when American management displays such greed and excess.

Union members want their employers to be profitable. But our members expect to share in the rewards of the profits that they help create. As we continue to refine labor-management relations in the years ahead, we look forward to more two-way cooperation with our employers. One of my objectives will be to encourage more CEOs of good union companies to muster the courage to speak out publicly in support of their union relationships. This is good business in every sense of the word.

I have long been troubled by the reluctance of most corporate leaders to break ranks with their peers. Even corporate leaders who have good relationships with their union counterparts sometimes feel that they have to remain silent when their colleagues bash unions. This keeps them in good favor with their friends in the boardroom.

A number of years ago, when labor was pushing legislation requiring a sixty-day notice for plant closings, I called then AT&T Chairman Jim Olsen who had given a speech opposing the bill. I asked why he decided to speak out publicly against the legislation when his own company was already required to follow a sixty-day notice of plant or office closing because it is written in our contracts.

Olsen told me that he opposed the bill simply because the rest of corporate America did not want it passed and he was going along. I am sure it occurred to him that this measure was one way to level the playing field for AT&T with its nonunion competitors who were able to lay off or shut down with no notice and little cost. But his desire not to stray from the "club" overrode what would have been a sound business decision.

This also occurred when President Clinton proposed national health care insurance. In 1994, the president asked me if I could get some of the telephone company CEOs to stand with him and other corporate executives, primarily from the auto industry, just on the employer-mandate section of his bill. Each of the CEOs turned me down, notwithstanding the fact that 25-30 percent of their health care costs are generated by workers who do not have employer-provided health care and use the hospital emergency rooms as their health care provider. The employer-mandate provision would go a long way to reduce their costs by requiring all employers to provide a minimum level of insurance.

I later wrote to each CEO saying that they apparently don't mind paying for those who have no insurance—so don't tell the union you have a problem at our next contract talks. They would rather not break with the Chamber of Commerce than do what was right for their companies.

The chairman of Ameritech, Richard Notebaert, however, co-operated with the Administration at the outset. He offered to do a health care rally featuring President Clinton. He and I flew with the president on Air Force One to Milwaukee, where each of us addressed employees and their families. The rally was broad-cast to all Ameritech locations in the five states and back to Washington where CWA was holding its legislative conference.

So, why did Mr. Notebaert decline the president's later request to support him on employer mandates? It is my understanding that the Republican members of Congress from Ameritech's five-state region were so infuriated at Mr. Notebaert's cooperation with the president that they threatened non-support of the telecommunications legislation that was under discussion in Congress. They were playing hardball.

A few years ago, Bruce Carswell, then-executive vice president of GTE, invited me to a meeting of the Labor Policy Association, an organization of vice presidents in charge of human resources for major companies. Bruce was the newly elected president of the LPA. I spoke about the things that I believed labor and management could work on together. I was most conciliatory. Later, Bruce told me that he was severely criticized by some in the LPA for inviting a union representative. I was the first labor leader ever invited to the Labor Policy Association, and probably the last.

There are many good union employers worthy of praise. Top executives who publicly stand up for labor include Bell Atlantic CEO Ivan Seidenberg, SBC CEO Ed Whitacre, and American Income Life Insurance Chairman and CEO Bernard Rapoport. There are others.

I am alarmed at what appears to be the emergence of a new class warfare against labor by the business community as a result of our resurgent economic and political influence. In December 1997, Thomas Donahue, the head of the U.S. Chamber of Commerce, called union leaders "thugs in blue suits" because the AFL-CIO had the nerve to publicize CEO salaries on its Executive Pay Watch web site. Another corporate leader, a member of the Business Roundtable, chided his colleagues for not recognizing the "threat" that labor poses to them in a fundraising letter to raise political dollars to "stop" us.

Conflict between labor and management is not inevitable. Workers and managers both have an interest in a viable, successful company. We live in a competitive world economy and ultimately job security depends on profitable and productive employers. At the same time, the new global economy will be increasingly unjust and inequitable unless the labor movement substantially increases its representation of the U.S. workforce. Organizing, with all of the expenses, frustrations and heartbreaks it can entail, remains labor's greatest challenge—and likewise, our greatest opportunity—as we look to the future.

CHAPTER SEVEN

Organizing

My union career began as an organizer, bringing my own bargaining unit into CWA over strong company opposition. Then, after three years of learning how to be an effective local union president, Joe Beirne offered me the opportunity to organize 18,000 New York Telephone plant workers, a task that was successful three years later. I worked literally seven days a week for those three years, until victory was ours. I have never stopped organizing.

Organizing requires total commitment. Nothing else matters.

During the New York Telephone campaign, the press reported that Joe Beirne, along with other union leaders, was planning to meet with Soviet Dictator Nikita Kruschev when he came to the U.S.

This was in the coldest of the Cold War days. The independent union that represented the New York Telephone workers had been labeling CWA as a left-wing union. In fact, some of our key supporters had filed a libel suit when they were called communists. So it didn't take long for two key in-plant organizers, Jack Lafferty and Lou Hennigan, to tell me what a setback that meeting would be for our campaign.

I called Joe Beirne and asked him not to meet with Kruschev. The degree of my commitment to the organizing campaign was clear—I, a temporary employee terribly intimidated by my president, still had the nerve to confront him over a major union policy issue.

Joe said: "You want me to give up a possible contribution that I could make to world peace for the sake of 18,000 new members?"

I responded: "I don't care about world peace." And at that moment, that's how I felt. I realized later that sometimes there are larger issues to consider. But back then all that mattered to me was getting those 18,000 workers in the union.

In the end, Joe did not meet with Kruschev. When the other labor leaders went and the meeting got favorable press, I heard about it from Joe.

Being an organizer has to rank in importance with your family. Without family support, it is difficult to maintain both responsibilities. Something will have to give. Understanding that, I have always urged new staff to put their family first, and never stay in the field overnight when there is a way to get home.

But it doesn't always work that way. I remember when my daughter Janice was about nine or ten and I sent her to her room. I can still see her at the top of the stairs, stamping her foot and saying: "Besides, you're never home when we need you."

I turned to my wife and started to say, "You told her that," but Florence interrupted me.

"Yes, she heard it from me. Because in family emergencies, you were almost always away."

They were right — and it hurt. Organizing can become all consuming. No matter how I tried to put family first, there were occasions, too many of them, when my family suffered. That's why I was incredibly lucky Florence was supportive of my union work and believed in it like I did.

Organizing is very tough work. It involves following up leads wherever they take you. I remember giving out handbills in front of telephone buildings in areas where we had little or no support. I took insults and was rebuffed. I was also shoved and subjected to other provocations. But I learned to take it all with a smile. Then, one day the first person took the leaflet and a few days later one or two workers talked to me. We finally had our opening.

Suffolk County was a real tough area for us. The President of Local 1174, Mike Marino, a coworker of mine in Mackay Radio, called to tell me that a telephone company friend of his was willing to meet us at Mike's home on Sunday. My colleague, Joe Volpe, and I drove one hundred miles in a blinding snow storm to make that meeting. As a result of our efforts, we finally got an in-plant organizer for Suffolk. I wonder whether we sold him on the merits of CWA, or whether it was our determination.

Organizing is not a job or a career. It is a mission. It takes dedication, a commitment to make everyone's life better. An organizer must have a fire in his or her belly. At times, the work is tougher than just about anything a person could do. But it is also the most rewarding.

An organizing campaign is a tremendous experience. You feel indescribable tension as the votes are counted. When the tally shows that you have won, it is one of the greatest feelings of accomplishment you can get as a trade unionist, only surpassed by the jubilation of the in-plant committee composed of men and women who labored for months, often longer, putting their jobs at risk as they fought for their basic right of selecting a union of their choice.

Not everybody makes a great organizer. Some people just don't have the desire or the tenacity, even though they are committed to the union and are very effective in other roles. And even good organizers get burned out. Ultimately, nearly every organizer reaches a point where he or she should move on to other things. So we have agreed with the CWA Staff Union that when vacancies occur in our districts, organizers have the right to seek those positions. If someone has been a successful organizer, I believe he or she can do just about anything else in the union.

We continually try to highlight our commitment to organizing. One of the first things I did after being elected president was to eliminate the post of Organizing Director and upgrade that responsibility to Assistant to the President.

As CWA enters the 21st century we must raise organizing to the highest levels of elected leadership, such as making it the responsibility of our executive vice president, as it was when I joined CWA.

Our goals in organizing are two fold. We need to expand the number of workers who are represented by unions. At the same time we need to organize in industries and sectors where we already have members to protect our standards. As our union grows, we also create more opportunities for diversity in staff positions.

For a long time, CWA's organizing strategy was prioritized this way:

A) Bell telephone workers;

B) Other telephone workers; and,

C) Other workers within the communications industry.

Now our priorities are:

A) Workers within our core industries – telecommunications, other media and communications, information services, public and health care;

B) Other targets of opportunity – basically employee groups we think we can organize because of varied circumstances; and,

C) Companies within our core industries that are actively antiunion.

As companies in our core industries start new lines of business, new subsidiaries or agree to joint ventures, our survival as an effective voice for workers is jeopardized if we permit the growing parts of the business to go nonunion. So growing within those companies will always be top priority.

In 1954, when my Mackay Radio colleagues and I first joined CWA, it was very difficult for many in the union to see it change from its historic status as a union solely for telephone workers. Joe Beirne didn't know where to put us. Because we were a national unit, with six local unions covering workers in New York, Washington, D.C., California, Hawaii and ship to shore stations on all U.S. coasts, Joe felt we were more closely matched with our AT&T Long Lines national unit.

At the outset of the 1955 Long Lines national unit conference, a motion was made to expel us. We were telegraph workers. We were "different." After heated debate, a majority of the locals finally voted to include us. It reminded me of the fellow in the hospital who got a telegram from his coworkers that read: By a vote of 18-16, we hope you get well.

I will always be grateful to two local presidents, Charles Gantz from Philadelphia and Helen Fowler from Louisville, who spoke out strongly for our inclusion. They had a vision of the future of CWA as a growing, diverse union.

Today, reaching out to all workers who need a union is fully accepted in CWA. But it took some pioneers with guts and vision to make it happen. That change in attitude showed me that this union could deliver for whoever joined. It doesn't matter what job one does.

In early 1987, then AFL-CIO President Lane Kirkland told me that the International Typographical Union (ITU) was being raided by the then-unaffiliated Teamsters Union. President Kirkland felt that if ITU members were not offered an additional choice, the Teamsters stood a good chance of winning. He did not want that to happen. He asked me to offer CWA as the other choice.

I had mixed emotions and expressed them to Lane. ITU is the oldest union in the country. It has a glorious and rich history. Its Washington, D.C. local is the oldest in the nation and its charter is in the Washington Monument. But it was a declining union. Years earlier, when confronted with new technology dramatically changing the composing room, ITU leaders made a wrong choice.

Instead of adopting a policy of demanding that printers be trained on the new technology, they opted for guaranteed lifetime jobs for the incumbents. Most newspaper publishers snapped at the offer. They were willing to grant lifetime jobs in exchange for their unilateral right to put the technology where they chose, keep it nonunion and pay lower wages. And, when the last printer left, so did the union.

President Kirkland had another vision, however. He said that technology and marketplace conditions would require most of the unions in the media industry to merge. He did not believe they would be able to get themselves together; that it was far more likely that they would come together under CWA's umbrella.

He was so right.

While it is true the ITU membership continues to slowly erode, we have changed the old position on technology and organizing victories have brought in new members. We have become the home for 10,000 members of the National Association of Broadcast Employees and Technicians and the 30,000 members of The Newspaper Guild. The latter helped us attract 2,200 members who work at Dow Jones and *The Wall Street Journal*.

These kinds of mergers tend to make our union more attractive to others. As a result of the affiliation of the Coalition of Law Enforcement Agents of Texas, we have established a Department of Public Safety which has led other police and correctional unions to discuss joint activities and merger with us.

Every single one of these mergers or partnerships makes us stronger in our communities and benefits all CWA members and the labor movement in general. As we expand our presence in the community, we become better known and better able to attract new members.

When we target a company for organizing, for example, we ask our members in the community where that employer is located if they know anybody who works there. You would be surprised how many positive responses we get. Nearly everybody is related to somebody or knows somebody, or knows somebody who knows somebody who works for the company. That is one way to make direct, friendly contact with workers you want to organize. And it is good community organizing.

That is why organizing is best done at the grassroots, by members on the front lines. We must continue to direct our organizing resources to the local level. Locals also need to make a financial commitment to organize new members. Local organizers and activists, working with the support of the national union, are in the best position to motivate workers in their communities.

We can't simply walk into a workplace and promise employees higher wages and better benefits if they just sign a card. Instead, we have to educate them as to why they need a union; that they need a collective voice to speak to the employer about their collective concerns. The union reaches out and makes contacts, building on whatever prounion sentiment we find in the workplace. We find leaders and give them the support and the tools they need to organize their own workplaces. We educate, empower and mobilize employees so they can organize themselves. That's the job of the organizer.

The successful organizer builds self confidence in the workers he or she is trying to organize. The organizer has to establish absolute credibility and trust, proving himself or herself to be knowledgeable, sincere and honest. He or she should always be straight with the workers, never lying to them or making promises that the union can't keep. Most importantly, the successful organizer listens to the workers and educates him or herself about their concerns and aspirations.

Sometimes it is wages, benefits, pensions – the standard worker interests that a union can advance. But often it is simple quality of work life issues, like abuse from supervisors or parking spaces or bathroom breaks. I remember one female worker in a Connecticut factory that we were organizing who asked if the union could stop her supervisor from announcing over the loudspeaker every time she went to the bathroom. She had been ill and was taking medication that caused her to visit the bathroom periodically. Her physician gave the company a letter explaining her condition. Nevertheless, the supervisor treated her without regard to dignity.

Of course, the big issues like wages are important. But they are often less tangible than quality of work life issues like the one I previously described. Every single worker has suffered some kind of indignity or inconvenience at the job, which the union can address.

Unfortunately, during organizing campaigns, the union and its supporters can run into obstacles – from management and other workers themselves. Fear is the primary factor in workers' resistance to bringing in the union. They are afraid that they will get put on a bad shift, or they will not get promoted, or will even get fired if they support the union. And those fears are often proved correct. Each year about 10,000 workers who become active in an organizing campaign get fired.

The labor law that is supposed to protect workers who want a union actually works against them. Say that during an organizing campaign, one or more worker leaders are fired. The company will take advantage of all the delaying tactics available to them, and it will usually take two to three years before the case is adjudicated. By then, in many cases, the organizing drive has been defeated, or severely crippled.

Even if the company loses, they suffer no real penalty when found guilty of violating the law. The worker is ordered reinstated with back pay, less what he or she may have earned elsewhere. Except for their legal costs, which the company sees as the price of doing business, they are back at status quo, while the union that sought to organize the workers has expended a great deal of precious resources.

We need to reform labor law, creating not only an expeditious process to achieve justice, but also establishing stiff penalties so that companies found guilty of labor violations are fined and damages are paid to the victims.

Once CWA makes a commitment to organize a workplace, we remain committed. Some campaigns can take as many as eight elections before we finally win. The workers must have total confidence in the union and faith in the organizer.

Organizing the public sector workers in New Jersey, for example, took a long time and a great deal of commitment. Ed Schultz had come over to CWA from American Federation of State, County and Municipal Employees (AFSCME) in the late 1960s and went to work organizing the public workers in that state. Ed demonstrated an ability to organize the public sector and really made organizing a priority. He recruited Larry Cohen, a 25-year old public worker activist, to help him. Six years later, Larry succeeded in heading up the drive that brought 36,000 New Jersey state workers into our union in four separate elections. We were opposed by AFSCME and the American Federation of Teachers (AFT).

The labor reporter for the *New York Times*, Bill Serrin—back when they used to have a labor reporter -- once called Larry the best organizer in the labor movement. Whether he is the best, I am not in position to judge. But he is certainly up there with the best. Larry has got the fire and determination necessary to be a great organizer. His enormous energy and commitment inspire the people around him.

Our success in New Jersey made CWA more effective in organizing other public sector workers, not only government administrative workers, but also public safety employees. The potential for growth in public safety is enormous, because this is an area where many states are attempting to privatize government services. People who had for a long time considered themselves civil servants suddenly find themselves working for a national corporation that is running a prison or a welfare system for profit. These employees need to be organized and represented.

As the workforce becomes more educated and more professional, their concerns change from those of industrial factory workers. For the most part, professional and technical workers already have wages and benefits above the average unorganized worker. They are more concerned about issues like child care, elder care and having a voice on the job. They also want help in career development, particularly educational opportunities.

CWA already has an impressive record organizing this new workforce. Early in our history, when we primarily represented Bell System employees, most of our members were in the business of information services. As we reached out to other workers in other parts of the communication industry and the public sector, we became very knowledgeable about the special needs of these employee groups. We were able to tell workers that we understood their concerns and had dealt with them before.

In 1980, when I was District One vice president, I got a call from a nurse in Buffalo. She asked if I could meet with a group of her colleagues who wanted to join a union. I went to Buffalo and met with six nurses who called themselves Nurses United. They wanted CWA to organize some nine hundred nurses at Buffalo General Hospital.

"Why did you pick CWA?" I asked. "I was in a hospital once, that's all I know about nurses."

"We see your locals here in western New York," one nurse told me. "They are doing great work. We like what we see. We want to be a part of that. You negotiate with some of the biggest companies in the country, so you should be able to negotiate with this hospital. And what you don't know about nursing, we'll teach you."

Well, we had a tough fight organizing Buffalo General. Then we had to strike to get our first contract. At the time, Mario Cuomo was governor of New York, and I asked him to send in the health commissioner, who threatened to shut the hospital down until we got the contract.

That nurse was Debbie Hayes. She was elected president of the new local union, Local 1168. She is a dynamic leader. Not only does she represent her members well, but she never stopped organizing. Now CWA represents some 5,000 nurses and other health care workers in the Buffalo region. We are the largest nurses' union in western New York.

Our success in Buffalo is a good model for broad-based community organizing. In addition to the telephone workers, we represent employees in the local newspaper and television station and the hospitals. When people in that community see telephone, media, hospital workers and other CWA members mobilizing together, they see them all as CWA members, not as media or hospital workers. That is the kind of presence we want to have in communities all over the country.

Organizing, however, will always be difficult because unions have never been fully accepted by the business community. Corporate opposition to union organizing today is more sophisticated than ever. When industrial trade unions first began organizing in the 1930s, management often times sent in thugs who didn't hesitate to resort to violence, terrorism, and even outright slaughter to bust the unions. Those tactics are not acceptable any more. Instead, the company, trying to keep the union out, will hire an antiunion law firm and "consultants" to wage a campaign against the workers' efforts to organize.

Employers call these activities their "right to free speech."

But think about this for a moment. A working person's ability to form and join a union—freedom of association—is a fundamental constitutional right. The law of the land. But today, workers frequently have to risk their job, their livelihood, perhaps even, everything they have, simply to exercise their constitutional right. The labor law that is supposed to protect workers who want a union actually works against them.

There are people around today who try to convince us that most workers don't want to belong to a union because "unions are outdated." To them I say, try this test: Go into any bank, walk up to any teller and say in a loud voice, "I'm from a union. I'm here to talk to you about organizing a union of your own in this bank. Can we meet for lunch?" Watch the expression on their face immediately change from cheerful to stone. See the blood drain from their cheekbones and the pupils in their eyes dilate as the body's protective system from fear goes into overdrive.

This situation would not exist if the law worked properly to protect and guarantee our labor rights. But fear and conflict created by management spread throughout the workplace whenever workers try to form a union. This fear and conflict is the primary factor in workers' resistance to forming their union. And those fears are too often proved correct.

According to the Dunlop Commission, a government advisory panel that issued an in-depth report on labor law in 1995, 32 percent of nonunion workers said they would vote for a union where they worked. Another 13 percent said they were undecided. Of the 55 percent who said they would vote against the union, 12 percent said that they would change their vote if management didn't actively oppose the union.

We often see management-inspired fear and conflict when

workers attempt to join CWA.

In November 1997, we won the largest organizing victory in the private sector since 1987 when 10,000 customer service and reservation agents employed by US Airways voted to join CWA. The workers built their union under the most difficult circumstances. Labor law in the airline industry is governed by the Railway Labor Act which covers the transportation industries. While the National Labor Relations Act may be weak, the Railway Labor Act is worse.

Under the Railway Labor Act, a mail ballot is conducted in which a majority of the entire union-eligible workforce must vote to form a union. There is no "No Union" vote on the ballot. In other words, every ballot not mailed in is considered a "no" vote against the union. In addition, part-time workers and laid-off workers going back years may be considered part of the eligible workforce allowed to vote. And the union isn't even given the names and addresses of the workers who are participating in the balloting.

Management holds all of the cards under the Railway Labor Act. It is amazing that workers have formed any unions in the airlines, rails and other transportation industries.

In the US Airways campaign, management waged a bitter fight against CWA supporters. At one point, the company even sent antiunion messages over the computer terminals to the workers. What was most incredible to me was that at the very time we were lamenting the very low turnout of voters in the 1996 presidential and congressional elections, US Airways management, in writing, urged their employees to destroy their ballots. Each destroyed ballot would be a vote against the union. The company committed numerous violations of law which we were able to document. We lost the first vote. Fortunately, justice prevailed and the government ordered a second ballot. In 1997, the workers got their union in the second vote by a solid margin. But what a struggle for them!

This campaign was also a model in total union organizing. Because US Airways has employees in one hundred and ten locations all over the country, we got our local unions involved in the organizing effort. A local would take responsibility for a nearby airport. We also asked that whenever our members traveled through airports, they take a minute or two to speak with employees at US Airways. We took every opportunity to talk to the US Airways workers about our union.

During the campaign, we were able to point out to airline employees that CWA had 160,000 members who worked in other companies who did exactly what they did. Our experience and understanding of their particular problems resonated with the US Airways employees.

Most airline reservation and customer service representatives do not have union representation and we have heard from many others about their interest in joining CWA. We expect more to join our ranks in the future.

In 1994, one hundred and seventy-five employees of La Conexion Familiar, a San Francisco telemarketing company wholly owned by Sprint Long Distance, tried to organize a union with help from CWA Local 9410. The predominantly female and Hispanic workforce earned about $7 an hour, compared with an average hourly wage of $15.83 for telephone workers in 1994. Among other injustices and indignities they had to suffer, the company discouraged workers from drinking water in order to cut down on the need for bathroom breaks, even though the job required them to talk on the phone all day.

The La Conexion workers filed a petition for union recognition with the support of 70 percent of the workforce. During the organizing campaign, Sprint management engaged in a pattern of harassment and intimidation. They spied on activists, punished organizing leaders, and constantly threatened to close down the facility if the workers voted for the union.

On July 14, 1994, less than a week before the election was scheduled, Sprint made good on their threats. Over the public address system, management announced that La Conexion was closed and that workers would be searched by security guards on their way out of the building. All of the workers were fired, the office was shut down, and the work was transferred to another location.

Following an investigation, the NLRB filed a complaint against Sprint, charging the company with more than fifty labor law violations. Sprint admitted to many of the charges, and a company vice president was fired for fabricating evidence submitted to the NLRB.

An administrative law judge found the company guilty of most of the charges, but the main charge, that Sprint illegally shut down the La Conexion office in order to thwart the union organizing campaign, was appealed. In December 1996, two and a half years after the incident, the NLRB found Sprint guilty of that final charge and ordered them to rehire the workers and to provide full back pay. Nearly a year later, the U.S. Court of Appeals overturned the labor board ruling, leaving the workers with nothing.

In 1996, Sprint had $14 billion in revenues and $1.2 billion in profits. Sprint annually spends hundreds of millions of dollars on

advertising, sponsoring sporting events, and other ploys to enhance their public image, but they did not want to pay their La
Conexion workers a decent living wage. For Sprint, it clearly paid
to routinely violate labor laws in order to keep its workers from
unionizing. Apparently, the company felt that the legal bills in-
curred by the La Conexion battle were a small price to pay to keep
the union out of their workplace.

Not only was Sprint able to keep La Conexion workers from
organizing, but their aggressive actions against the organizing
campaign had a chilling effect on the organizing efforts of other
Sprint workers and sent a message that the company would do
anything—even break the law—to remain nonunion. Unfortunately, this attitude is not unique in the corporate world. In fact,
it is becoming the norm.

The La Conexion case inspired a study by researchers at
Cornell University to measure the impact of the threat of plant
and office closings on worker organization drives. The study found
that in fully one-half of all organizing campaigns, as well as in 18
percent of first contract negotiations, employers threaten to close
their facilities, and they follow through on the threat 12 percent
of the time. This represents a much larger increase in shutdown
threats than found in earlier studies. The result, the researchers
concluded, is that the possibility of an organizing drive's success
is reduced from 60 percent to 40 percent when the employer
threatens to close the facility.

I could tell numerous stories like this. Obviously, it takes a
lot of courage for workers in America to stand up for their rights
and become active in forming a union. The CWA organizer is
there for them, talks to them honestly about the possible consequences, and provides as much support as the union can. These

workers could get fired, but if they do, the union will try to get them jobs elsewhere. If the campaign suffers a setback, the union isn't just going to disappear. We are and will be there for them—no matter how long it takes. Just as we were at US Airways. Just as we remain for the workers at La Conexion. Just as we will be for any worker and their family who desires to be a part of the CWA family.

At the 20th century's end, there is no reasonable hope of changing the nation's labor laws now or in the foreseeable future to effectively prevent management-inspired intimidation, harassment, fear and conflict when their employees try to exercise their union rights. Unions must search for other methods and techniques to help workers.

We can start with ourselves first.

It is hard enough to fight management during organizing battles. We should not make organizing even more difficult by fighting with each other. Over the years, CWA and the International Brotherhood of Electrical Workers (IBEW) fought bitter and divisive battles within the telephone industry that lasted well into the 1970s. Finally, Joe Beirne proposed a compromise on organizing new facilities that AT&T was building. The compromise was accepted and neither CWA nor the IBEW has opposed each other since.

With 90 percent of the private sector workforce unorganized, it is absolutely unacceptable for unions to be battling one another in any enterprise. But this still happens today. CWA completely supports AFL-CIO President John Sweeney's efforts to encourage multi-union campaigns and to bring unions together so that we will work more closely with each other on organizing unorganized workers.

Business is certainly working together in cities and states across the nation to prevent workers from forming unions. A couple of years ago, CWA District Nine had its district meeting in Reno, Nevada. At the same time, the Carpenters' Union was organizing the Hilton Hotel. The Carpenters asked us to stay at the Hilton and help with the campaign. We helped on both fronts, and I remember standing outside at five in the morning passing out handbills along with other CWA volunteers.

The Hilton arranged their hotel personnel in such a manner that all of the workers on the CWA floors were already union supporters. That way, we didn't have ready access to the maids or room service workers whom the company figured we might be able to convince to change their vote. That is how sophisticated the antiunion campaign can be.

As it turned out, the Carpenters lost the election. Afterward, I asked the union's general counsel why Hilton had opposed the union. This was very unusual for a company that is not known for being antiunion. The counsel told me that the Reno Chamber of Commerce had met with the Hilton people and told them that Reno was a nonunion town. If Hilton wanted to operate in this community, the Chamber said, they would have to be nonunion.

After hearing this, I began to see the need for the labor movement to target the Renos of this country. We could have ten to fifteen unions go into a location and tell the Chamber of Commerce: "We are going to organize this city and we are going to be here for however long it takes. So you better get used to us."

That's what the labor movement is doing now in Las Vegas. The hotel and restaurant workers union has done a heroic job of making every Las Vegas operation a union workplace. And the building trades are working together to see that all new construction is a union job. This effort needs to be expanded among greater numbers of workers with a particular emphasis on the professional, technical and administrative occupations.

The time has come for labor to make the same Las Vegas type commitment to bring the union message into all the Silicon Valleys in America. CWA is prepared to lead that campaign. If unions are successful in organizing workers in such circumstances, we can then show the local business people the value that a union workforce brings to their employers and the community.

Because of the weaknesses in labor law, CWA is pursuing a strategy of avoiding the National Labor Relations Board whenever possible. We seek instead to take fear and conflict out of union organizing. We use the existing leverage and power we have—while we still have it—to win guarantees from our employers that management will not interfere with union organizing campaigns and to voluntarily accept union representation when a majority of the workers sign union authorization cards.

Our definition of management neutrality is absolute noninvolvement and noninterference by management when workers begin to discuss their interest in forming a union. We won this kind of strong management noninterference and card check recognition agreement with SBC in 1997 and use that language as a model for all of our contracts.

We simply cannot allow our employers to establish new nonunion lines of business or subsidiaries that threaten the jobs of our current members. Management noninterference and card check or quickly scheduled non-NLRB consent elections are an industrywide strategy for union organizing that expands the job security for our members by creating more opportunities for them to be reassigned or transferred to other union jobs. At the same time, we demonstrate the link to our members that a strong union is in their direct interest.

The workforce is changing and so too must union organizing. Today, women make up 37 percent of the union workforce, a higher percentage than at any time in labor history. That percentage will assuredly grow. The vast majority of new workers we will have to organize will be women and minorities. Women and minority workers have different experiences, different pressures and different concerns, in life and on the job. We need organizers who are attuned to their needs, who literally speak their languages.

We also have to reach out to young workers. In 1994, fewer than 8 percent of employed men under the age of twenty-five were members of labor unions. As the older and more proportionately organized workforce retires, we need to replace those workers and grow even more. If labor unions are not able to organize young workers, the future will not be very bright.

We need to reach out to new workers and organize wherever we see opportunities. After a long history of CWAs being a Bell System union, I have a simple test to know when we are truly diversified. It will be when every District Vice President can no longer refer to "the company" with everyone knowing precisely which company was being referred to. That will not happen until each district has enough sizeable bargaining units representing workers outside of the phone company. We are moving rapidly in that direction. This will add power to every one of our bargaining units.

CWA has been enormously successful in expanding the scope of our membership. Forty percent of our members are in bargaining units that were not unionized twenty years ago. I like to tell people today that it does not matter what company you work for—CWA can be your union. Today millions of workers perform basically the same tasks of information handling, customer service, computer and data operation, and other job responsibilities that our members have been performing for years. Unlike the traditional industrial unions organized around workers who helped create a single product line, in the new labor workforce the product is irrelevant. What matters is that the union understands and can help workers cope with the challenges and changes they face every day.

The work of organizing is never finished. As long as there are nonunion workers out there, we have a job to do.

Addressing our legislative conference in 1997, Vice President Al Gore closed by leading the chant:

"Early to bed.

"Early to rise.

"Work like hell.

"And organize."

I couldn't have said it any better.

CHAPTER EIGHT

Collective Bargaining

ollective bargaining is the heart of modern labor organizations. It is the best vehicle and most powerful economic strategy for workers to get a fairer share of the wealth they help create.

I have been directly involved in the collective bargaining process since the beginning of my career in CWA. When my unit at Mackay Radio joined the union in 1954, we started bargaining with a management that was determined to test and provoke us.

I learned a great deal during that first bargaining session. My mentor was Mike Mignon, a fellow radio operator who had been fired by American Cable because of his organizing efforts, and then went on to become a CWA staff representative. Mike was instrumental in helping organize us and then bargain our first contract. Mike taught me the importance of patience at the bargaining table. During the first month of bargaining, when it was clear the company wanted us to walk out, he sat at the table reading aloud from the works of Oliver Wendell Holmes. We didn't go anywhere and neither did the company.

Negotiations dragged on for many months. But through patience, intelligence and member support, we not only won a decent first contract, but also earned management's respect. That experience showed me how to bargain effectively.

In our next bargaining session, we negotiated what I believe was the first automation clause in American labor history. Manual radio and cable operators who were replaced by new technology (the teletype machine), were able to retain their job titles and pay rates. The company had the right, however, to assign them to do other work. No layoffs or downgrades were permitted. It was a lifetime job guarantee that also preserved the union.

We were later successful in changing the ITT pension plan so that the company could no longer force workers to retire before their Social Security benefits took effect.

Today, the collective bargaining process has opened exciting new innovations in the delivery of quality health care, child and senior care and worker empowerment and education while still improving wages, pensions and working conditions. Grievance and arbitration systems protect workers against injustices and hold supervisors accountable. In a nonunion shop, no matter what rules an employer may establish, they are unilateral and can be changed at any time. Establishing contractual obligations through collective bargaining agreements is the only way to guarantee that employers do not break their promises or go back on their words. Collective bargaining brings discipline into the workplace for both workers and management.

Not only does collective bargaining raise the living standards of union workers, but it truly is a rising tide that lifts all boats. When we negotiate a contract that includes decent wages and benefits, we raise the standard for everyone else as well. Unfortunately, it also cuts the other way. If union membership and union power diminish, labor's strength at the bargaining table also declines, and that affects everyone, not just union workers.

History is instructive as to how collective bargaining is directly linked to decent living standards for working Americans.

By passing the Wagner Act in 1935, Congress declared that it was the policy of the United States government to encourage collective bargaining in the workplace. The Wagner Act made it possible for unions to successfully organize on a mass scale and gave them legal protections once they were organized. It helped bring about a historic period of union growth when workers could form unions free of management harassment and threats.

In the years that followed, the collective bargaining process helped society adjust through periods of prosperity and recession by providing steadily increasing wages, benefits, safety and dignity for workers while at the same time creating stability for corporate management. Collective bargaining was part of the social compact between labor and management, and workers prospered as a result.

There is a direct correlation between the decline in real wages over the past 25 years and the lower percentage of workers who participate in collective bargaining.

During the Clinton Administration, America experienced sustained economic growth. The Dow Jones soared to record heights. Corporations posted ever-higher profits. Unemployment was low and productivity high. But most working families have not been able to share in the benefits of this economic expansion. Fewer of them are represented by unions and thus do not participate in the collective bargaining process. If union membership continues to decline and unions lose more power at the bargaining table, the income gap between the rich and working families will continue to widen. That is how important collective bargaining is, not just to working families, but to American society as a whole.

Collective bargaining provides the flexibility to respond quickly to changes in the workplace and address the concerns of workers. In today's workplace, for example, employees face new health threats such as carpal tunnel syndrome and other stress-related illnesses and injuries. Lacking government standards, these issues must be dealt with through collective bargaining.

In 1993, I was chairman of the AFL-CIO Safety and Health Committee. One of my first responsibilities was to participate in a Workers' Memorial Day event at AFL-CIO headquarters which was held to highlight the dangers in America's workplaces. Secretary of Labor Reich joined me and, in a closed one-hour session, we met with the families of workers who were hurt or killed on the job.

It was heartrending to hear the stories of the aftermath of workplace accidents, injuries and deaths. A little girl brought tears to our eyes when she asked why her Daddy died. I know that Bob Reich walked away with a new understanding of how dangerous the workplace can be and how important collective bargaining was in protecting workers' health and safety. But, of even greater importance is the role government must play in enforcing good safety laws. Every accident that we heard about could have been avoided. If the collective bargaining process does not cover more workers and government does not aggressively enforce the law, there will only be more carnage in the workplace.

Since CWA is so broad-based, collective bargaining for us is a year-round activity. We have contracts with more than 2,000 employers and even when we are not at the bargaining table, we are always looking to use union leverage to win battles away from the table for our members. For example, CWA recently did a stress survey of our members. I sent the final report to the CEO of every major employer. I told them that we could work on trying to solve the problem of worker stress now, or we could wait until bargaining. Not a single company disputed our findings. All of them recognized that certain jobs are extraordinarily stressful. Reducing stress would benefit the companies as well – in less absenteeism, higher morale and better efficiency. In many places committees were established to find answers to the problems.

After divestiture, as I discussed previously, CWA's collective bargaining relationship with the telecommunications industry changed dramatically. Before the breakup, we thought we had some tough bargaining sessions, but, in retrospect, they were nothing compared to negotiations after divestiture. Notwithstanding, we were able to resist virtually all of the concession demands made by management of the former Bell System companies. We successfully blocked health care cost shifting for both active employees and retirees, because we made clear that health care cuts were a strike issue.

In 1986, the first bargaining after divestiture, we went on strike against NYNEX over health care cost shifting. The strike was a week old when Jan Pierce, my successor as District One vice president, asked if, in light of my relationship with NYNEX CEO Bud Staley, could I try and resolve it.

It took me overnight to locate Mr. Staley. I finally found him on Sunday at the racetrack in Saratoga, where he did not yet appear too concerned about the strike. We sat in a hotel lobby and I told him neither of us needed this strike; that it could last a long time. To give him a way out, I proposed that we postpone any cost shifting to the first day of the next contract (in 1989), so that we would have another opportunity to bargain. He agreed and the strike ended.

But there was a problem and I didn't find out about it until April, 1989, when I was attending a meeting of the board of directors of The Alliance for Employee Growth and Development. I got a call from a Mr. Donovan, who identified himself as NYNEX's vice president in charge of labor relations. I had never met him. He told me that CWA leaders were telling our members not to sign the payroll card to pay for their health care. I called Vice President Pierce and quickly learned we had a severe problem. The payroll card provided for shifting of health care costs to our members.

After our meeting in Saratoga, Bud Staley apparently had gone back to his side and told them that there would be cost shifting the first day of the new contract. Staley thought we had already given in on cost shifting, when the language of our agreement clearly stated that we would protect health care for the 1986 contract and then negotiate it further in the 1989 bargaining session. Not being a labor relations practitioner, Staley missed the implications of "another opportunity to bargain."

An honest misunderstanding led to a seventeen-week strike during which one of our members was killed on the picket line. But the strike ended in total victory for the union. We not only protected health care for our NYNEX workers, but also sent a message to every other employer that our members are willing to fight to protect their health care. Further, where they are willing to fight, this union will spend every last dollar it has or can borrow to support them. We paid $32 million in strike benefits to CWA members, $16 million of which was borrowed.

After the strike, Bell Atlantic's Labor Relations Vice President Ed Grogan came to my office. He told me that the labor relations executives of the other companies "were wondering where NYNEX was coming from."

He told me that when the Bell and AT&T labor people met after the 1986 round, the NYNEX representative announced they were going to get health care cost shifting from CWA in 1989.

"Not with that language," said Roy Howard of BellSouth.

All the other company people agreed with Mr. Howard, Grogan said.

I confirmed these events with Roy Howard and Dave Hudson, AT&T's representative, who also was at that meeting. I then called Bud Staley, who had retired. I reported my conversations to him and suggested if his guy at that meeting hadn't come back and said, "maybe we have a problem," he should be fired.

Sixty thousand families—40,000 CWA and 20,000 IBEW — had gone on strike and sacrificed because one man didn't know his job. Finding out how it all happened was important to me to maintain my credibility with the industry.

Staley's successor, Bill Ferguson, was a man of integrity and courage. He took over as CEO during the strike. The company had already cut the health care benefits of management employees and was operating under the assumption that it would do the same to the union workforce as well. Bill Ferguson and I responded to an offer from Jack Stafford, CEO of American Home Products and a member of the NYNEX Board, to use his Senior Vice President, Joe Bock, and their good offices to reach a settlement.

We met at noon on Sunday at American Home Products in New York and reached agreement at about 5:00 a.m. Bill Ferguson had the guts to withdraw their demands that enabled an accommodation to be reached. We were very grateful to Jack Stafford for his interest, Joe Bock for his nudging and humor and Mac Lovell, who was appointed by Secretary of Labor Elizabeth Dole to mediate the dispute, for his patience and skill.

While we have been successful at the bargaining table, we do not always get what we want. At times, we give in to a company demand, but it is not so much a concession as a trade. I have always taken the position that there is a price for everything, except basic union values. If the company wants a change in the contract, and that change is not a matter of principle for the union, I might agree to it, so long as I am able to gain something in return for our members. The elected members of our bargaining committees and our members, of course, have the final say because they must live under the contract.

Some unions in other industries have had to capitulate to management demands simply to save jobs, often responding to a management threat to move the work elsewhere. Management has seized upon their ability to move capital, jobs and technology to force some unions to make painful concessions. Once a union is on the defensive, management gains a tremendous advantage at the bargaining table. The only sure way to reverse this concessionary trend is to organize and increase union power. As long as unions see themselves as primarily or even exclusively service organizations for their existing members, they will be less successful in attracting new workers. This results in less clout at the bargaining table. The organizing mode is the winning way.

Despite the tougher bargaining climate of the past few years, CWA has been able to secure significant wage gains for its members, keeping them above the national average. And we have broken new ground in protecting workers from job displacement due to technological change and other factors. Contracts covering the vast majority of our membership now provide for joint discussion of the impact of new technology, as well as training and retraining opportunities so that workers can remain employed and even move up into better jobs.

As a result of a growing membership and aggressive tactics, we are not afraid to experiment with new approaches to bargaining. For example, companies are coming to us and asking to bargain early. In 1998, seven months before our contracts with SBC expired, covering some 76,000 workers at Southwestern Bell, Pacific Bell and Nevada Bell, we received a request from the company to go into early bargaining. Before we agreed to early bargaining, I had the CWA vice presidents in charge of negotiations, Ben Turn and Anthony Bixler, determine where the company stood on several of our key issues and what their demands were going to be. The first criterion was that the company could not have any negative proposals.

As it turned out, SBC appeared close to where we wanted to come out and agreed not to make any proposals that we felt were harmful to our members. The company was more anxious to have three years of labor peace during a time when competition was bound to get even fiercer. We realized that it was a strategic gain for CWA to enter into early negotiations not only to insure a good contract for our members, but also to set standards for the other companies we would be negotiating with later on. Once SBC agreed to certain demands, it would be easier for us to hold other companies to similar stipulations.

After about three weeks of bargaining, we negotiated a new thirty-two month contract, that was unquestionably the best agreement in the industry since 1983. Both sides were winners. SBC Chairman Ed Whitacre got what he wanted: stable and positive labor relations through March 2001. At the same time, 76,000 CWA members will enjoy higher wages, better benefits and improved employment security. SBC also saved millions of dollars in not having to prepare contingency plans to counter a CWA mobilization campaign.

Since we negotiate with many different companies in diverse regions within the same industry, it is vital that our first contract be a good one. We can then hold the other companies' feet to the fire. In previous years, while we have been largely successful in imposing the parameters of our strongest agreements, we have agreed to modifications to fit the local conditions.

There is no question that bargaining has gotten tougher with all of our employers in today's highly competitive environment. But competitive pressures do not necessarily have to raise the level of confrontation in labor-management relations. In fact, with GTE, changes in the marketplace are moving us toward a much more cooperative relationship. GTE is one of the largest telecommunications companies in the nation and has expanded greatly through acquisition of smaller companies. Therefore, with a few exceptions, their properties are scattered across the country, with bargaining units ranging in size from fifty to 12,000.

GTE maintains a corporate labor relations staff at its Irvine, Texas headquarters. It also has labor relations staff at each of the companies around the country. Historically, negotiations at each company were virtually conducted as if that company were a stand-alone enterprise. The same issues are always fought at each company, although there is also some centralized control from headquarters. Serious differences of opinion could usually be worked out if the president of CWA and the executive vice president of GTE (who is in charge of labor relations) spoke together. If we agreed, the executive vice president had to "persuade" the local company to go along.

As you might expect, the process of negotiating with GTE

companies is cumbersome, confrontational and extremely costly. We generally spend as much time negotiating for a 200-member unit as for a 12,000-member unit.

As a result, CWA, IBEW and GTE are in an ongoing process of developing one of the most far-reaching cooperative ventures in the telecommunications industry. This program will certainly change labor relations at GTE and very well might evolve into first regional bargaining and hopefully even national bargaining with the company in the years ahead. We are forging this venture at a time when other industries are moving away from national bargaining or pattern-setting agreements. As a result of our developing relationship with GTE and its progressive management under CEO Charles Lee, President Kent Foster and Executive Vice President Randy MacDonald, we know we will be able to build a solid cooperative partnership with a mutually beneficial plan for the future.

The GTE venture began with a full-day meeting attended by our vice presidents, representatives of the IBEW, and the top GTE management in December 1996. At the end of the day, as we were struggling to reduce to one paragraph all we had discussed, Tom White, GTE's Telops President whom I had never met prior to that meeting, wrote a note and passed it to me.

"How does this work for you?" he asked.

The note read: "GTE – A Wall-to-Wall union company and leader in the marketplace."

"Where do I sign?" I responded.

Those few words epitomized a win-win relationship between the union and its employer. Good relationships are key to successful collective bargaining, and in the future our relationships must change from confrontational to cooperative. While we will continue to be tough when circumstances require, we also must forge more mutually beneficial relationships with our employers.

Pounding the table and storming out of the room might make one feel good, but does it accomplish anything for our members? It is easy to be adversarial, the tough job is to find common ground with an employer. That is when the union representative faces a true test of leadership.

We have been working to improve our relations with GTE for more than a decade. In 1986, I participated with the district leadership and local bargaining committee in the first days of talks with GTE of Southern California. The contract covered some 20,000 CWA members. The president of the company, David Anderson, also joined his bargaining team as we shook hands to officially open bargaining. It was the first time that a president of CWA and a president of a GTE company participated in the opening of contract negotiations.

I repeated the same scene a couple of months later on the eve of the start of negotiations with GTE-Southwest, a unit then of more than 6,000 members. I arranged for the union and company bargaining committees to have dinner together in San Angelo, Texas. The company president Buddy Langley and I joined them. It was an evening of general discussion and breaking the ice. The next morning Mr. Langley and I were at the table to kick things off. Again, we set another first for the company and the union, and those relationships set a tone for a successful round of bargaining with GTE.

For any union where the president is involved in the collective bargaining process, his or her role is to develop a personal relationship with the CEO and other top officers who can make things happen. Any such relationship has to be built on trust and credibility. The president has to be able to call the CEO minutes before a strike deadline and come to an agreement. The CEO has to know that the president can deliver on a promise.

My role with every CEO is to have the kind of personal relationship that can make the process work better and help our members. Our relationship today with SBC is a good example of this.

Ed Whitacre, CEO of SBC, has demonstrated through concrete action that he fully accepts our union in a partnership and understands our need to grow with his company. He agreed to total management non-interference and card check recognition in all SBC units and new ventures, both now and in the future. But it was not one-sided just for the union. We assisted SBC at the state regulatory and federal levels and played an important role in an international acquisition.

In consultation with our union leaders in SBC, I invited Ed to speak to our 1997 convention. I wanted to thank him by having him recognized by our 2,000 convention delegates who gave him a standing ovation. I also wanted to send a message to all of our other CEOs that they, too, could be singled out for this honor.

The experience also showed the maturity of our local union representatives. The last time a CEO had been invited to a CWA convention was 1956 when Joe Beirne invited the CEO of Southern Bell to speak to our delegates. It was a year after the bitter seventy-two day strike against Southern Bell and Joe felt this may be the way to put the past behind us. But our delegates were not ready and it was a disaster.

Whatever I do and wherever I may be, I always ask: How can I advance the interests of the union and its members? Here is just one example: CWA represents NABET workers employed by NBC, which is owned by General Electric. I was invited to the

White House State Dinner given for British Prime Minister Tony Blair, and Jack Welch, CEO of GE, was also in attendance. I had never met him, so I asked AFL-CIO President John Sweeney to introduce me to Jack and we had a conversation.

Now if a major problem arises with NBC, I feel comfortable to phone Jack Welch. This is an avenue that may not have been available to us before.

In 1997, CWA won the bargaining rights for 10,000 customer service representatives at US Airways. I had never met US Airways CEO Stephen Wolfe and wanted to find someone to introduce me rather than going in cold to meet him. I discovered that Bell Atlantic CEO Ray Smith sits on the US Airways Board. A call to Ray was all it took for a threesome lunch, with Ray making a tactful retreat and leaving Mr. Wolfe and me to have some quality time together.

In February 1998, I attended a fund raiser for the Kennedy Library. I do not particularly like putting on my black tie, but when I do, I look around the room to see if I can conduct some CWA business. I spotted Mr. Wolfe and we talked about the then ongoing negotiations for our first contract. In fact, I was surprised with his knowledge of the status of bargaining. We both agreed that a contract settlement was in our mutual interests and left the door open to each other to push it to a conclusion if necessary.

You don't have to become personal friends with the person who sits across the table from you, but sometimes friendships do develop. I had that kind of relationship in the late 1960s with Ricks Litell, head of labor relations for New York Telephone. Ricks and I often had dinner together. Sometimes, he would call me the next morning and ask what we had agreed on the night before. I would have it written down on a cocktail napkin, because I am always working. He never questioned my notes.

When Ricks was transferred to vice president for human resources at Michigan Bell, New York Telephone gave him a farewell party. Someone asked his wife what she thought about the move. She said, "The only thing I'm worried about is that I may not be able to sleep because I'm used to sleeping with a telephone cord across my chest, since either Ricks is calling Morty or Morty is calling Ricks in the middle of the night."

I was able to build the same relationship with Ricks' successor Ray Williams, who went on to become vice president at AT&T. Ray was instrumental in helping us establish The Alliance for Employee Growth and Development. I always told Ray that he had paid his dues having been in the chair on the other side of the table in the 218 day-strike against New York Telephone in 1971-72.

Besides giving more credibility to the process, a personal relationship with your negotiating partner enables you to overcome obstacles that otherwise might be insurmountable. A personal relationship makes it much harder for either one of you to walk out of the room, and there is much more honest and direct communication. You never have to wonder what that person really means.

There is also a potential danger of the relationship growing too strong. That was on the verge of happening with Litell and me. We both agreed that we may be in danger of losing our objectivity and needed to stay a little more at arms length.

It is more difficult to build strong relationships these days. As a result of the breakup of the Bell System and the downsizing that followed, virtually everybody from "the good old days" is gone. Today in some telecommunications companies, management has little or no experience in labor relations. This has been difficult, not only for them, but for us as well.

It is much better when you know the players. Shortly after I became president of CWA, the president of Southern Bell invited me to attend a management seminar in Florida. Ben Porch was our vice president for District 3 at the time, representing workers in nine southern states. Not only did Ben know every single manager at that conference, a couple of hundred people, by their first names, but he also knew many of their fathers. That was an enormous strength when it came to collective bargaining. Unfortunately, it has all but disappeared in today's volatile and ever-changing environment.

It is not easy to break in or train management with whom there is no personal relationship and who are not labor savvy. Recently, I got in a telephone argument with one of the top officers of Bell Atlantic. I guess I got a little loud. He said, "This is why our people would rather not have the union."

"You're missing the point," I replied. "If you expect me to agree with everything you say in order for the management to accept the right of their employees to have a union, it will never happen and we will never build a relationship."

Collective bargaining is a constantly evolving process. Today, our collective bargaining relationships make it possible for CWA to better prepare its members for the vast changes ahead. It was only through the collective bargaining agreements that some of our most innovative programs and partnerships were developed. The Workplace of the Future, the Alliance for Employee Growth and Development, Pathways to the Future, The Next Step and so many other programs came out of collective bargaining. Without the union's support of such programs and demanding them during bargaining, they would not exist.

Not only does collective bargaining make the workplace fairer, it also makes the workforce more productive. A workforce under a collective bargaining agreement is both more stable and more flexible, and it is in the best long-term interests of employers to have their employees protected by such agreements.

Built into our Workplace of the Future process with AT&T is the potential for a living contract, in which bargaining agents are given the authority by union members and the corporation to amend the contract to meet changing market needs, without having to wait as long as three years for the next contract. This helps make the employer more competitive, but at the same time could be a big plus for union workers.

Living contracts would radically change the process of collective bargaining. Instead of locking ourselves into contracts every three years, we would continually adjust the contract according to circumstances. This would only be possible if certain issues, like health care and pensions, were taken off the table entirely. Instead of just being an opportunity for the company to seek modifications during a temporary downturn, a living contract would give the union the ability to ask for increased wages and profit-sharing during a time of growth. If we are to agree to living contracts, it would have to be a two-way street and the trust would have to exist at the very highest possible level.

We are a long way from living contracts as a general rule, but I feel they are the future of collective bargaining. In a hypercompetitive economy that is undergoing such rapid and radical change, we need to be more flexible while at the same time protecting basic union values and worker needs. I believe we have a great opportunity to reach a living contract with AT&T through Workplace of the Future.

Contrary to what some so-called experts might think, collective bargaining is not an anachronism that is ill-suited to a changing economy. Collective bargaining is democracy and the free market system in its purest form: worker representatives, democratically elected by their co-workers, sit and negotiate as equals with their employers over the fair market value of their worth to the enterprise. Collective bargaining provides the opportunity for management and labor to work out problems of mutual concern while insuring that workers receive just and fair compensation. Through collective bargaining, management and the union can experiment with new forms of work organization and compensation that will have the support of the workforce because they have a voice in setting the terms of the agreement. Collective bargaining institutionalizes these changes by taking them out of the realm of one person's leadership and offers the opportunity for management and labor to build on past success.

No wonder, as noted labor economist Paula Voos pointed out recently, that the most innovative and lasting work practices today are taking place in union workplaces. In order to have a significant impact on the American economy, we need to reverse the antiunion policies that have undermined collective bargaining. We need a positive national agenda to promote worker representation through collective bargaining agreements. One way to help bring about these much needed changes is for the labor movement to increase its political clout, along with a substantial increase in union organization and political activity.

CHAPTER NINE

Political Action

"I wouldn't want to be a member of CWA in New Jersey this morning."

This was the threat levied by one of the highest ranking Republican leaders of the New Jersey legislature the morning after Election Day 1997.

The Republican Party was hell bent to punish CWA and its 60,000 members who work for New Jersey state, county and city governments because we were out front in almost defeating Governor Christie Whitman.

A quick and outraged response by CWA brought an equally quick retreat by the GOP leader. While the apology was public, the intent to silence unions and their members remains.

We face greater political challenges than at any time in our recent history. In the 1930s labor unions received enthusiastic support from FDR and his New Deal. The Wagner Act guaranteed the right of workers to organize and established legal mechanisms to protect that right.

Bill Clinton and Al Gore are the first president and vice president to publicly announce their support for organized labor in more than thirty years, but the antiunion forces among the right wing in Congress and corporate America have still been somewhat successful in passing their agenda. When they were not entirely successful, they put the labor movement on the defensive. We spent much of our energies fighting laws and initiatives that would hurt working families, rather than advancing programs that would help them.

CWA is active politically at every level of our union. We have members in every Congressional district, and in several states are not only the largest union, but also the most politically effective. We try to use our power to advance the needs and aspirations of not only CWA members, but working families everywhere.

When CWA endorses a candidate, we do not just look at the party label. We look at the candidate's record, votes and personal support for the issues that directly impact the economic and social well-being of our members.

Over the years, the overwhelming majority of candidates supported by CWA have been Democrats. That is because the values and the principles of Democratic Party candidates tend to reflect our concerns, our values and our principles. Since the New Deal, labor unions have historically been aligned with the Democratic Party. The Democrats have been more responsive to the concerns of working families. However close we are to the Democrats, we are allies, not captives.

In the 1994 elections, when the Republicans took over Congress, shockwaves thundered throughout the labor movement. As labor leaders, we were stunned. Obviously, we had not done a good enough job informing our members about where the candidates stood on the issues.

We had made a valiant effort to block the North American Free Trade Agreement in 1993, which was supported by President Clinton when the Democrats controlled the House and Senate. That treaty was passed over our objections and much of or-

September 17, 1954 — Mike Mignon, center left, signs our CWA first contract at American Cable and Radio (formerly Mackay Radio). This was the culmination of three years of hard work. It was worth every second of the effort.

Being sworn in as the new officers of CWA Local 1172 by District Director Mary Hanscom. *From left*, Bob Murphy, Secretary-Treasurer, Mickey Gill, Vice President, and Morton Bahr, President. My first question was "what do I do now?"

With Joe Volpe, the organizer who introduced me to CWA.

When our son Dan was born in 1946, I was at sea. This is the actual telegram I sent to Florence. In 1996, we celebrated Dan's 50th birthday with his wife Marilyn, left, and daughters Heather, upper right, and Shelley.

With our daughter Janice.

Our daughter Janice, center, with her daughters Alison, left, and Nikki.

Then and now — With
Florence on our honeymoon
in 1945 and today.

MORTY AND MEL - AGES 3 and 6
BAHR

The aspiring catcher — Growing
up in Brooklyn, I was consumed
with baseball.

With my brother Mel, left. We were ages 3 and 6
respectively when this was taken.

Two great leaders — CWA's founding President Joe Beirne, right, and our second President, Glenn Watts, who was Secretary-Treasurer when this photo was taken in 1973.

A dream realized — On January 16, 1974, Joe Beirne left his sick bed to announce that the Bell System had agreed to National Bargaining with CWA, a goal Joe had pursued since 1947.

A couple of "Happy Warriors." — Vice President Hubert H. Humphrey and CWA President Joe Beirne.

Nate Fine Photo

A rare photograph — taken in 1973 — of Glenn Watts, left, Joe Beirne and me. Glenn and I have been privileged to follow in Joe's footsteps and have the opportunity to serve as President of CWA.

Sam Reiss Photo

With the late Senator Robert Kennedy.

My first CWA Convention in 1954, with President Beirne and District Director Mary Hanscom.

Solidarity in the streets. At a rally to support CWA bargaining with AT&T in 1995 (top); a huge rally for the striking Detroit newspaper workers in 1996 (center); and, marching in support of the United Federation of Teachers in New York back in 1968.

Florence and I with President Bill Clinton.

With Vice President Al Gore at the White House.

With First Lady Hillary
Rodham Clinton at a CWA
Convention.

It was an honor to shake hands with South African President Nelson Mandela.
That's former AFL-CIO President Lane Kirkland in the background.

Senator Ted Kennedy at our home in New York.

With Governor Mario Cuomo of
New York.

ganized labor approached the 1994 elections quite demoralized and upset with the Democrats.

As surveys later showed, our membership reflected this disillusionment with the Democratic Party and its leaders. Our members voted 41 percent Democrat, 40 percent Republican (only 20 percent are registered Republicans), and 19 percent voted for a third party or did not identify their choice. That pretty much reflected the country at large. Even worse, union turnout was historically low. Union families constituted only 16 percent of the total votes cast in the 1994 elections. The first two years of Republican control of Congress reawakened labor's political consciousness. We fought hard to win a minimum wage increase over the opposition of Republican congressional leaders and defended Medicare, Medicaid, and other essential social programs from drastic cutbacks.

A heightened sense of political awareness spread throughout labor and nearly every union concentrated on directly communicating with their members about the important issues at stake in the 1996 elections. CWA, for example, produced three union-wide membership mailings on the key issues during this period, a first for our union. Union members voted overwhelmingly for President Clinton and labor-endorsed candidates for Congress, and, indeed, almost recaptured the Congress. Just as important, our turnout numbers dramatically increased to 24 percent of the total vote and we had a significant influence in tempering the right wing, antiunion ideological leanings of Congress.

Obviously few, if any, of us in the labor movement want to be tied to a single political party. The ideal political climate for organized labor would be to have an ideological majority in both parties that is pro-worker.

We support Republicans who promote the issues that are important to working families. We refuse to support Democrats who do not have a good record on those issues. There are usually about twenty Republican members of the House who vote with us on critical labor issues. Unfortunately, the leadership of the Republican Party is controlled by antiunion special interest groups that not only oppose us on labor and broader economic and social issues, but are also opposed to our very existence. With encouragement from some in the business community, they want to see us severely weakened, or taken out of the political mix altogether.

This is unfortunate, because some 20 percent of CWA members are Republicans and they are effectively disenfranchised from their own party. I would like to see them get involved with their party and try to bring it back to the center. If they are rebuffed, they may want to reconsider why they are members of a party that, for example, believes it is okay for a worker to be permanently replaced when on a lawful strike.

Former Secretary of Labor Robert Reich is a very decent person. He became one of this nation's most effective Labor Secretaries. He cared about workers and he was willing to learn.

However, it took Bob Reich about three years on the job before he mentioned the word "union" in public. I continually urged Bob to talk about collective bargaining as one of the best ways for workers to share the fruits of their own labor. Finally, Bob sent me a copy of a speech he had given, along with a note saying that he had taken my advice. I sent him a note back saying that his speech was great, but "when will you make it in this country?" The speech he sent me praising unions and collective bargaining had been delivered at the International Labor Organization in Geneva.

Ultimately, Bob Reich did begin talking about collective bargaining in America, but the fact that he took so long points to a curious double standard in our current political climate. Every other cabinet secretary can advance the cause of their constituencies. It is perfectly all right for the Department of Commerce to carry water for corporate America. The Secretary for Veteran Affairs is unashamedly "pro-veteran." The Secretary of Education is never blamed for being "pro-education." But for some reason the Labor Department cannot be seen as being "pro-labor."

That double standard is hypocritical and unacceptable. I believe that the Secretary of Labor should, whenever possible, make the case that the collective bargaining process is good for America and be an advocate for the rights and aspirations of working families. The Secretary of Labor should never apologize for supporting workers and their unions wholeheartedly.

The double standard does not just occur at the cabinet level, it exists throughout our political system. Some politicians are perfectly happy to accept contributions from labor PACs but then are reluctant to publicly voice their support of us when they are not speaking to labor audiences.

We need public officials to use the bully pulpit on behalf of union workers and the collective bargaining process. Decent wages, benefits, working conditions and employment security are not partisan issues. Mario Cuomo proved that in 1982 during his primary campaign for governor of New York against New York City Mayor Ed Koch. Koch had the support of virtually every special interest group. Cuomo had the support of New York labor. Koch attacked Cuomo early and hard for being a captive of organized labor. Throughout the attacks, Mario refused to run for cover. Instead, he responded by paraphrasing Samuel

Gompers: "If being for better schools, health care for everyone, safe communities and decent jobs means that I am a captive of the unions, then I am proud to stand with them." Koch retreated because his polls told him that Cuomo was right. It is up to us in the labor movement to make that case to the American people and our political leaders.

Joe Beirne taught me that politics is about relationships. He had close relationships with Harry Truman, John F. Kennedy, Lyndon Johnson, Robert Wagner, Jr., and other political heavyweights of the time. Joe also taught me to always make sure that when you support a political leader, be very clear about what you want in return.

Joe called me in 1974 after reading in the paper that I was the labor chair for Howard Samuels' campaign for governor of New York. He said: "When I call you the morning after Election Day and ask you what CWA got out of your work, don't tell me good government." Samuels didn't get elected, but other politicians whom we supported did.

In 1972, the New Jersey County welfare workers, whom we had organized, had not received a pay increase for several years because Governor Cahill had refused to sign our negotiated contracts. I met with Judge Brendan Byrne, who was running for governor at the time and told him, "CWA is going to bust our butts for you this election. One thing we really need is for you to sign these welfare board contracts."

Byrne got elected and signed the contracts. It also was the right thing to do.

Now that I am president of CWA, I have a wide variety of political relationships. One week's political activity in February 1997 shows how those relationships allow me to advance the cause of our members and other working families. This was my schedule.

Meeting with Vice President Al Gore, Air Force II—In early 1997, Archbishop Desmond Tutu invited Vice President Gore to begin a dialogue with South African trade union leaders. The vice president asked me to go with him to Durban, where the meeting was held. Having the vice president as a captive audience for twenty hours on Air Force II was a great opportunity to discuss the issues with him.

AFL-CIO Executive Council Meeting, Los Angeles—Following our meeting, the vice president dropped me off in Capetown where I caught a flight to Los Angeles on Saturday night in order to attend the AFL-CIO Executive Council meeting.

Meeting with John Podesta, White House Deputy Chief of Staff—The first day of the Executive Council meeting, I had a private session with the new Deputy Chief of Staff to the President John Podesta. I told him of our continuing concern about President Clinton's favorable mention of Sprint in his State of the Union address. I reminded John of Sprint's conduct at La Conexion Familiar and that Sprint was the nation's top labor law violator. I also called to his attention recent decisions by the FCC which created an unfair playing field in telecommunications that could cause us to lose hundreds of jobs in the regional operating companies before they were permitted to enter long distance. I received a commitment that both issues would be dealt with in the White House.

Lunch with Minority Leader Gephardt—The next day, I had a private lunch with House Minority Leader Dick Gephardt. I briefed him about our concerns with the FCC docket, the problems associated with the Welfare Reform Act and the undemocratic procedures of the Railway Labor Act where we received some 4,000 votes from US Airways workers and lost the election because management convinced enough workers to trash their ballots.

Meeting with Presidential Economic Advisor Sperling—The following day, I had a private meeting with President Clinton's top economic advisor, Gene Sperling, who serves as chairman of the President's National Economic Council and has enormous influence on public policy. We talked about strategy to implement President Clinton's policy to disqualify companies like Sprint from getting government contracts, as well as my work as Chair of the Commission for a Nation of Lifelong Learners.

Vice President Gore attacks Sprint—Vice President Gore flew from South Africa to Los Angeles later that week and addressed the AFL-CIO Executive Council. He had met with eight workers, all of whom were either discharged, suspended or otherwise harassed by their employers simply because they wanted a union. Among the eight was a CWA supporter who had been fired when Sprint closed its La Conexion Familiar subsidiary in San Francisco one week before the election. In his remarks, the vice president singled out Sprint to express his outrage at their corporate conduct. He went on to lay out the Clinton-Gore policy that would expose such outrageous corporate lawbreakers to the public. The vice president also promised that the president would veto the TEAM Act and any amendment to the Fair Labor Standards Act that would destroy the forty-hour work week. Further, Gore said that the president promised to issue an Executive Order that would bar lawbreakers such as Sprint from getting federal contracts. He assured us that we would hear both him and the president publicly supporting unions and the collective bargaining process. His statements caused the U.S. Chamber of Commerce to attack the president for "tilting the labor-management relationship towards the unions." We are nowhere near a level playing field, much less a favorable tilt to labor.

Private meeting with Vice President Gore—Following his remarks, I gave the vice president a chronological history of Sprint's labor abuses, from the closing of La Conexion in July 1994 to the NLRB's far-reaching indictment of the company. I told the vice president that we were shocked to hear the president, in his State of the Union speech, commend Sprint for hiring a few welfare recipients. Vice President Gore expressed his regrets and later that evening privately told me that he would personally talk to the president about Sprint.

I do not pretend that this is a normal work week for me, by any stretch. But it does illustrate how important it is to build positive relationships. Because of our relationship, Vice President Gore has not met with the telecommunications companies as a group without me present. It is always made clear that I am there as the representative of their employees.

We can't take for granted that every sympathetic politician has a first-hand understanding of our issues. Both Vice President Gore and Secretary Reich were "pro-worker." Neither had the slightest idea of what workers went through to organize a union in their workplaces until they began to meet with victims and listened to their stories.

Gore and Reich were outraged. The vice president said he was certain that if the public was aware of such outrageous corporate activity, as they were exposed to, things would change.

"Public exposure," Gore said, "will make such conduct unacceptable."

Our goals are to elect men and women who are sympathetic

to the concerns and aspirations of working families. We want people to have decent jobs, to live in decent houses in safe communities, to be able to raise healthy families, to be able to educate their children and start them out in life, and finally be able to retire comfortably, with a secure pension and healthcare.

Within these broad goals are many specific issues that both CWA and the labor movement in general are concerned about and try to address at all levels of political activity. Our political agenda includes these issues.

Job Safety—Every day an average of seventeen American workers are killed due to workplace injuries, and thousands are injured by toxic substances, explosions and other hazards. That adds up to 7,000 workers killed, 60,000 permanently disabled, and six million injured on the job every year.

Congress needs to pass legislation reforming the federal Occupational Safety and Health Act. OSHA inspections are way down. Even the most hazardous job situations given the highest priority can expect to see an OSHA inspector only an average of once every thirteen years. And more than seven million American public employees remain outside the coverage of federal law. Of these state and local government workers, some 1,600 are killed on the job each year. Congress needs to strengthen OSHA to stop the carnage in America's workplaces.

Health Care—America's health care crisis has not gone away. It has gotten much worse. Some 41 million Americans are now without health insurance, and by the year 2005 that number will have risen to 47 million, nearly one fifth of the population. America will spend more on health care than any of our foreign competitors, and at the same time, many of our citizens will receive lower quality care. Compared to other industrialized nations, the U.S. ranks eleventh in maternal mortality, eighth in life expectancy and twenty-second in infant mortality.

The reason for these tragic statistics — the United States is the only industrialized nation without some form of national health care now that South Africa, under President Mandela's leadership, provides universal coverage.

The health care crisis also creates pressures on the insured, particularly union families. Every time we go to the bargaining table, we face management demands to cut health care benefits and shift costs onto our workers and retirees, those least able to afford it. These pressures increase as more workers lose their health care.

Incremental reforms to our health care system are nothing more than band aids. At the very least, we need to pass legislation requiring all employers to provide some minimum level of health care to their employees. Senator Edward Kennedy has introduced legislation to do this and his tenacity assures us that it will become law someday.

Family Medical Leave—The first bill that President Clinton signed into law was the Family Medical Leave Act. Even after this legislation took effect, the United States still has the worst maternity benefits for working women of any industrialized nation. We need to expand and improve upon the Family Medical Leave Act so that working mothers have sufficient time to care for their newborn infants before returning to work.

Minimum Wage—We support increases in the minimum wage,

even though no CWA members and few union workers earn the minimum wage. A higher minimum wage boosts the earnings of other low-wage workers. It helps working families. A full-time job at the present minimum wage is not enough to bring a family of four above the poverty line. And despite Republican rhetoric to the contrary, increasing the minimum wage does not eliminate jobs or slow job creation.

The minimum wage is particularly important now that so many states are trying to pay workfare employees less than the minimum as they move off welfare. If workfare is actually going to provide an opportunity for welfare recipients to get back into the workforce, then it must provide a decent living wage.

Privatization—Privatization in the public sector is comparable to subcontracting in the private sector. Work that union members are now doing would be performed by nonunion employees. CWA opposes privatization not only because we represent public sector workers whose jobs might be threatened, but also because we believe it is a bad idea. We have amassed numerous case studies of private companies that make any kind of promise to perform work cheaper, better and faster just to win a government contract. And they rarely do, with taxpayers left holding the bill.

In 1997, when Governor Bush tried to privatize welfare in Texas, CWA fought him, not only because it was bad public policy, but also because privatization would have spread like wildfire across the nation. We formed a coalition with AFSCME and the Service Employees International Union (SEIU) who had members in other states at risk. AFL-CIO President Sweeney gave us solid support. We estimated that some 250,000 people could have been at risk. CWA itself had about 12,000 jobs immediately at stake in Texas. We mobilized politically in Texas, using grassroots actions as well as radio and print advertising campaigns to explain why privatization was wrong.

At the same time we worked closely with the White House, convincing the president not to grant waivers that the state needed in order to implement the program. By generating so much heat at the local level, we made it easier for the president to deny the waivers. The privatization effort in Texas was defeated. But it is certain to arise again and again.

Permanent Replacement Workers—Workers cannot be fired when they go on strike, but they can be "replaced" by an employer without hope of returning to work. What's the difference? It is grossly unfair to allow companies to use replacement workers to break strikes. Even the threat of hiring replacements by management can chill the collective bargaining process.

CWA is not concerned with replacement workers in our contracts with large employers. But, it has hit us hard in other industries, such as publishing and newspapers. The nation badly needs national legislation banning the hiring of permanent replacement workers to end this despicable practice.

Company Unions—The same people who support the company's right to replace us when we go on strike want to give our employers the right to establish company unions. The pioneers who founded CWA know very well the dangers posed by company unions. CWA was born out of the company unions that existed in the old Bell System that were declared illegal after the Wagner Act was passed in 1935 and declared constitutional in 1938.

Now the Republicans want to turn back the clock sixty years and return to the days of company unions. They have introduced a bill deceptively named the Teamwork for Employees and Management Act (TEAM Act). This legislation would permit employers to set up and control worker committees. The Team bill provides management with a ready means of undermining unions and destabilizing bargaining relationships. Not only do we need to defeat the Team bill and other company union proposals, but we need to strengthen existing labor laws so that such so-called unions are clearly prohibited.

Right-to-Work Legislation—Right wing Republicans would like to pass a national right-to-work law. Not a single Democratic lawmaker supports this proposal. We call it the right to work for less law, because in the twenty states that already have right-to-work laws, workers have the lowest wages in the country. "Right to work" has nothing to do with the right of a worker to have a job. It merely allows states to pass laws that prevent companies and unions from negotiating contract provisions that require all workers who benefit from the agreement to pay their fair share for contract enforcement and administration. It is just a veiled attempt to weaken unions financially.

Fair Labor Standards—Antiunion forces in Congress also want to amend the Fair Labor Standards Act that now requires employers to pay time and a half after forty hours work during a single week. Instead, they want employers to be able to pay straight time if their employee does not work more than eighty hours in two weeks. That means employers could schedule workers to work long hours one week, and then part time the next week. As long as they do not exceed eighty hours in two weeks, they will not have to pay time and a half. This, say some Republican lawmakers, is family friendly legislation.

Labor Law Reform—We need to change existing labor law to reduce conflict and to strengthen the right of workers to join and form their own union. We need dramatic changes in the way workers can organize, how unfair labor practices are handled and to make corporate law violators pay with more than a slap on the wrist. Regardless of the political climate, labor law reform remains at the top of labor's political agenda.

Paycheck Protection Act—How can you be opposed to a bill named the Paycheck Protection Act? Very easily, if you know what the bill is truly about—eliminating labor unions from political activity. Unable to pass this legislation in the Congress, antiunion groups have begun to introduce similar legislation and/or referenda in the states. While this measure has already been deemed unconstitutional in a court of law, legal challenges will likely continue for the foreseeable future. Even if we defeat these proposals—as we did in California in June of 1998—the right wing has succeeded in distracting our attention and resources at a time when we need to advance our own legislative agenda.

Campaign Finance Reform—The only way to really fix campaign financing would be to establish public funding of political campaigns. But public financing is not a realistic possibility in today's political climate. So we have to work within the system as it exists. This means that labor unions have to raise and give money through their political action committees, and any change in the law which would diminish their ability to do so should apply to all players, not just labor unions.

Intellectual Property Rights and Privacy Issues—The convergence of information technologies has created new opportunities for employers to exploit the work of performers, artists, writers, newspaper reporters, broadcasters and other creative talents. Digital technology allows any piece of information to be transmitted electronically anywhere in the world. Our members deserve protection for their intellectual property rights. At the same time, all Americans must be assured of their privacy as the new electronic technologies are developed. CWA is in the forefront of protecting individual and privacy rights in the Information Age.

The political process offers union members the avenue by which we can promote our agenda for America. We encourage our members to get involved in the political process by participating in the local AFL-CIO and CWA COPE activities.

In the new century, our political goals must also change. We have to start running and electing union members and union supporters for local community boards, school boards, city and town councils.

The AFL-CIO plans to have at least 2,000 union member candidates ready to run for office in the near future. They will be mostly at the local level and well trained. The success of this plan is critical to the futures of working families in the next millennium.

By getting our people elected to government offices, labor can directly influence state and local politics. Not only would they influence the policies of local governments, but they would also be involved in the processes of primaries and general elections for higher level political office.

We also have to take full advantage of the power of mobilizing union households in elections. Most elections are determined by voter turnout. As Americans have become more cynical about government, voter participation has dropped to record lows. This gives the labor movement extraordinary opportunities.

Getting out the union family vote is important to electing pro-worker candidates. When they don't come out, as in 1994, our candidates lose. When they vote in large numbers, as in 1996, our candidates win.

But labor's greatest opportunity is in midterm elections, when general voter participation drops sharply. When the unions get a big union family vote to the polls, our endorsed candidates generally win.

We saw this play out in the 1998 special Congressional election in California. The California Republican Party saw an opportunity to pick up a House seat and went all out to beat Democratic candidate Lois Capps. Capps ran on a working family platform. Forty-five percent of the eligible voters went to the polls. But, with the unions working hard at the grassroots, 55 percent of union families came out and pulled Ms. Capps to a solid victory.

It is this kind of mobilization in key states and Congressional districts that must be a regular part of every union leader's responsibility. We know that by educating, motivating and getting union families to the polls, we can prevent the election of an anti-union president and Newt Gingrich-type leadership in Congress. We know what that would mean to workers and this nation.

The alternative? We would soon find the U.S. on a par with the Congo in terms of workers' rights and working conditions. It would not take long for company-paid healthcare to disappear. The public school system would be decimated, if not entirely destroyed. Pension plans would be eliminated, and workers would have to rely on 401K plans in times of economic volatility. The Social Security system would be privatized and run by Wall Street. A national right-to-work bill would be passed. The right wing would control appointments to the National Labor Relations Board and the federal courts, filling those positions with antiunion ideologues. We would have a Department of Labor that was no different from the Department of Commerce, merely serving the interests of corporate America. Collective bargaining would be a mockery.

It can happen here.

I don't want any CWA member to wake up on the Wednesday morning after any election day saying, "I could have done more." In 1996, a right wing candidate won a Congressional seat in Pennsylvania by eighty-four votes. The difference between Newt Gingrich and Dick Gephardt being House speaker that year came down to 11,000 votes in eleven districts, or an average of 1,000 votes a district.

In a democracy, every single person makes a difference. If you don't think your voice counts, and you don't use it, then the special interests will make all the decisions for you, and pretty soon apathy will become a self-fulfilling prophecy—your voice will not be heard because you didn't use it.

What I recommend to the members of our union could apply to all who believe government should work for its people. Every CWA member and family member should be registered to vote. Keep informed on all issues that affect you as a worker, parent and citizen. Know about all of your elected officials. Keep score as to who is standing up for workers and who is for the big corporations, regardless of party affiliation.

CWA members can rely, as can most union members, on your local and national union to do its best to give factual, nonpartisan information so you can make up your own mind. We recommend, but you make the decision.

The other thing I would ask our members to do is be more active in the local union and in the community. Successful political action begins with community-based unionism. Our ties to the community strengthen our ability to raise money, elect our friends and influence legislation. It enables us to resist attacks designed to weaken us. The union has to be active in all of the communities where its members live.

CHAPTER TEN

Community Service

In 1987, CWA was organizing the workers at Stephen Austin University in Nagodoches, Texas. These workers made the beds, did the cleaning, and cooked and served the food for the students. They were mostly female, both African-American and Hispanic, and were severely exploited by the contractor who was given the contract to provide these services to the school administration.

Working through Jobs with Justice, we planned a march through the town of Nagodoches and since the school got state funds, we were entitled to hold a rally on the campus itself. The night before our march and rally, I came to town and met with the workers in a little shack that served as their organizing headquarters. Some of the women spoke up about their working conditions and the treatment they regularly received. After recounting a series of indignities, one worker asked, "Do I look like an animal? Why do they treat me like one?" It was a highly emotional experience.

The next morning we started marching about two miles out of town. As we marched through an African-American neighborhood, the sidewalks were crowded with residents who supported the organizing campaign and what the union stood for, but would not join our march line. I went over to ask some of them why they did not join us. I will never forget one elderly man who said,

"We're all supporting you, but you'll be gone tomorrow and we have to live here." I thought, the civil rights revolution missed this town.

When we arrived at the campus, the students came out. They were all white. The administration also came out, and they were also all white. In my remarks, I asked them: "Consider for a moment the way you treat those who make life more comfortable here for the students, who are right now cleaning their rooms and serving their food. What kind of future leaders of America are you turning out here?"

We won the election and wound up representing the university workers.

While it was important for CWA to organize the workers at Stephen Austin University, it was equally important for us to help the entire community of Nagodoches. When organized labor does not have a presence in a community, there is often no effective voice for working families. The same kind of indignity and injustice that we see in unorganized workforces can also be present in unorganized communities. Unionized workplaces clearly contribute to healthier communities.

The rally and winning the election in Nagodoches were just the beginning. In 1997, we negotiated our fourth contract. We have a high level of membership and over the last ten years have won a number of arbitrations. They've expanded the unit and received recognition for the part-time employees. There are four members in the unit that are active on a committee that is expanding the local by trying to organize other university workers. They are a living example of what a trade union brings to even the most out of the way workplace.

CWA has traditionally held a broad vision of our role as a

labor union. We believe in contributing more to our members' lives than just an extra dollar in their paycheck. While it is important for them to have good working conditions, it is equally important for them to have good living conditions, for themselves, their families, and for other members of their communities. Economically healthy, safe and family-oriented communities are as important to the labor movement as collective bargaining power.

Community service has been a hallmark of CWA activism from the beginning. Joe Beirne called CWA "The community minded union." He believed that the union's responsibilities were not contained within the four walls of the enterprise where its members worked. Rather, since those members were whole people, we had to be a whole union. We had to not just make work better, but make life better.

This commitment to community shows clearly that labor unions are not special interest groups. The goal of the labor movement is a rising standard of living and better quality of life for everyone, not just our members. Every special interest group, whether it is the Business Roundtable, the American Trial Lawyers Association, the tobacco and drug companies, or the NRA, looks after its own narrow interest, not the community at large.

Compare that to labor's activism on behalf of working families and society as a whole. The Family Medical Leave Act, which was largely supported by labor, affects every working family in America. So does the Fair Labor Standards Act, the minimum wage, workers' compensation, unemployment insurance, pension portability, and a variety of other labor-inspired initiatives.

But the contributions that the labor movement makes are rarely acknowledged by the community at large. Instead, union leaders are frequently attacked as union bosses or thugs. Critics ask: what does labor want?—as if our agenda were somehow suspicious or excessive. When Samuel Gompers was asked that question, the popular and erroneous notion is that he greedily responded: "More."

What the history books fail to give us is the full quote that Gompers said:

"What does labor want? We want more schoolhouses and less jails, more books and less arsenals, more learning and less vice, more constant work and less crime, more leisure and less greed, more justice and less revenge."

Labor's heritage of community service goes back to the beginnings of the trade union movement in this country. Indeed, many unions first began as beneficial associations, providing help to the unemployed, the widowed, and the orphaned. That philosophy has continued through organized labor's history and has taken hold at the grassroots level.

CWA members believe that they have a special responsibility to be involved in their communities and are involved in a wide number of community services. They are active in such programs as Boy Scouts and Girl Scouts, Big Brother and Big Sister, the United Way, as well as less well-known services like mentoring adults and children, helping their local schools, and being involved in local politics.

At the national level, we are and always have been deeply committed to the United Way and its predecessor agencies, including the Community Chest and the Red Feather. Joe Beirne was chairman of the CIO Community Services Committee, and then after merger with the AFL, George Meany appointed him chairman of the AFL-CIO Community Services Committee. Joe was the first, and still only, chairman of the board of the United Way of America who came from the labor movement. In 1974, the United Way established the Joseph A. Beirne Community Services Award, which is presented each year to a union member for his or her outstanding contributions as a volunteer worker in the community.

CWA presidents have continued Joe's commitment to the United Way. Glenn Watts served on the United Way board, and I have been active in the United Way movement for some twenty-five years, including twelve years service as vice chairman of the Board of Governors. I was very proud to receive the Joe Beirne Award myself in 1997. I now serve on the Board of the United Way International. And CWA Secretary-Treasurer Barbara Easterling took my place on the United Way Board of Governors.

From its beginning, United Way's mission has been to bring the community together to meet human needs. The power of United Way to harness all of the many different resources in our communities behind a common vision is evident every day, where each of us lives and works. As part of our commitment to the United Way, we are very active in blood drives, and CWA members are among the most effective in organizing and carrying out those drives. Without such efforts, there would be a serious donor-blood shortage throughout the country.

Since the telephone industry is a service which everyone uses, and its employees have a great deal of direct contact with the public, phone companies have always been very conscious of their public image. Because of this, they have always been involved in community service.

We have engaged in many different community service projects with our telecommunications employers. The Pioneers are retired phone company workers, many of them former CWA members, who are involved in public service in the communities where they live.

One of the projects we are most proud of involved putting telephones in VA hospitals all across the country. In 1994, a CWA shop steward brought to my attention that there were no bedside telephones for patients in any of the 172 VA hospitals nationwide. Patients had to use pay phones in the hospital lobbies, and many of them couldn't get to those phones without assistance. Those who were bedridden had to wait for a nurse to bring a telephone in a seventy-five pound cart to the bedside. Friends and relatives often found it difficult to get through to patients, because the nurse was not always available to pick up the phone in the cart.

Frank Dosio, a CWA shop steward in Poughkeepsie, New York, is a Vietnam veteran who had a friend in the Veteran's Hospital in Albany. Dosio knew how difficult it was for his friend's family to contact him from their home in Buffalo, three hundred miles away. When his friend died, Dosio decided to do something about the problem. He went to his employer, NYNEX, and got them to contribute a piece of voice-activated telephone equipment. Then he got CWA technicians to volunteer their labor and install the equipment in the Castle Point, New York VA hospital. More than 200 members of Local 1120 donated labor to wire that first hospital.

Governor Mario Cuomo came down for the first phone call, which was made by a quadriplegic who was finally able to place a call to his father just by the sound of his voice. The Castle Point project took five months to complete.

Now called PT Phone Home and headed by Frank Dosio and coordinated with CWA Executive Vice President M. E. Nichols, the project's initial success quickly got other locals and their employers involved all across the country.

Following the completion of the Castle Point VA hospital

project, Dosio was nominated for the Points of Light Award given for outstanding volunteerism. He won the award and was honored by President Clinton at the White House. In presenting Frank with his award, the president said: "This shows how one person with a vision and passion can make things happen."

I spent a few minutes with the president after the ceremonies ended. I told him that the project had escalated not only to put a voice-activated telephone in each hospital, but to also put a phone at every bedside. I asked the president to make the first phone calls from the Oval Office to three hospitals that would be ready for telephone service on Memorial Day, 1994. He readily agreed.

I got to the White House early the morning of Memorial Day and went to the basement where CWA members were setting up the circuits. The emotion in that room was strong. Tough technicians were wiping their eyes. When they set up the circuits for test purposes, they had the vets from the three hospitals talk to each other. Veterans of World War II, Korea and Vietnam, for the first time, were able to talk to one another from different hospitals.

The president was just great in talking to the veteran who was chosen by his peers in each hospital. Vice President Gore also spoke. I had tipped him off that we had placed phones at all 600 beds. He concluded by saying: "I better get off the phone because I know there are 600 of you poised to call home."

The Oval Office was packed with press, print, radio and TV. As big a human interest story as this was, not a single reporter asked a question about the program. The only news reports were in the local papers, where the VA hospitals were. The national media ignored the story.

CWA and IBEW locals, along with volunteers from the Pioneers, donated the labor to install the phone systems all across the country. CWA employers at AT&T, GTE, NYNEX, Bell Atlantic, BellSouth, Ameritech, SBS, US WEST and Pacific Telesis all came through. Our employers all cooperated and deserve a share of the credit. By Fall 1996, not only were all 172 VA hospitals completed, but there was a telephone at every bedside.

The success of the project was celebrated in the Oval Office with the president making the first call to a World War II nurse in the final hospital. Fittingly, Frank Dosio was present and received the appropriate recognition from the President. Don Reed, executive vice president of NYNEX and a decorated Vietnam Veteran, represented the industry.

The Veterans Administration reported that our project saved the taxpayers millions of dollars. Of the people the program actually serves, few, if any, are CWA members. We did not do it just because it helped our members. We did it because it was the right thing to do. Now Frank Dosio is busy working to create remote classrooms for veterans and disabled children. PT Phone Home has already built computer classrooms in fifty VA hospitals and wired several schools with video phones, television screens and computers so that sick children can keep up with their school work from home. Disabled vets are learning computer literacy and will become productive, functioning men and women.

Our volunteer work also extends to community emergencies. In the past few years, Americans have suffered an unusually high number of national disasters. We have seen floods, hurricanes, tornadoes, earthquakes, and other disasters that have killed and injured people, destroyed their homes and livelihoods, damaged their communities and made it difficult to rebuild their lives. CWA members, like the population in general, have been directly affected by some of these catastrophes. And we also help whenever we can.

CWA established a National Disaster Assistance Fund, which is funded by voluntary donations and administered by Executive Vice President M.E. Nichols. When a district vice president determines that an area in his or her district is a disaster area, we move quickly to assist our members directly with financial assistance.

CWA and its members were very active following the devastating Hurricane Andrew in Southern Florida in 1992. Our employers donated computers and free telephone service. While working with the Florida AFL-CIO, our members created and manned a resource center to assist people to get home repairs from reputable contractors. The labor movement was able to ensure that the repair and construction work was not only performed by union workers, but was done honestly and at a fair price.

Along with the IBEW, we man the Hurricane Disaster Network along the southern Atlantic coast. Most of the disaster stations are either in CWA and IBEW local union offices, and the station operators work closely with the Red Cross and AFL-CIO Community Services Department.

CWA is also very involved with the Elizabeth Glaser Pediatric AIDS Foundation. In 1990, we met Elizabeth Glaser, whose husband was the television actor and director Paul Michael Glaser. Elizabeth had contracted the HIV virus through an intravenous procedure, and ultimately passed the disease on to her two children. Her daughter died early on.

Secretary-Treasurer Barbara Easterling invited Elizabeth Glaser to address our convention in Toronto. She gave a very moving speech.

"Let me tell you what your support means to us," she told the delegates. "First of all, it means that you are not like the rest of the world, because you have said that we must care and we will care about this. We will help to make it better.

"There is no way that you can ever really know how important that is until you stand in my shoes. And I hope that none of you ever has to stand in my shoes, because as you say that you care, others will see that they can care, too. You will set an example for a world that has still not decided to care."

Elizabeth later said that my embracing her on that dais was her coming out of the closet. She felt we were all now part of her extended family. And we were.

Normally when we take up a collection at a convention, we realize about $5,000. This collection topped $30,000.

Elizabeth went on to testify before Congress and addressed the Democratic National Convention in 1996. She carried the fight for increased research for pediatric AIDS.

Since our Toronto convention, CWA has remained involved with the Pediatric AIDS Foundation, making it our major charity. We are the largest single giver on a continuing basis, donating more than $2,500,000. We also made a commitment of $100,000 a year to support the research of one scientist for five years.

Elizabeth Glaser died in 1994. But she will forever be remembered as a courageous member of the CWA family.

Our experience with Elizabeth and Pediatric AIDS has also taught me a lesson in sensitivity. Pediatric AIDS is an easy charity for people to support, because the victims are children, all of whom contracted the disease either in utero or through a blood transfusion. If you have a heart, it goes out to these kids.

After discussing Pediatric AIDS in my *CWA News* column, I received a letter from a member who was HIV positive. He questioned my sensitivity, saying that in my warm support of Pediatric

AIDS, I might have given the wrong message—that we can put our arms around sick children because they are innocent, but adults who contract the virus are somehow to blame for their condition.

That letter really opened my eyes. I never wanted to send that message, and as a result of my being educated by a member, we are now working hard with Paul Glaser and others in Pediatric AIDS to make the case that AIDS is a health issue, not a moral issue.

CWA and its members are also involved in the Special Olympics, the Diabetes Foundation, Muscular Dystrophy and many other charities. During our 1994 convention, we helped launch a national bone marrow donor drive. As part of that drive, Kathy Champion, executive secretary to our Secretary-Treasurer, Barbara Easterling, was found to be a match for a thirty-four year old woman who needed a bone marrow transplant. Kathy donated bone marrow at George Washington Hospital and the recipient was able to survive, but unfortunately, for only a short time. It was a one-in-a-million match that only happened because of CWA's activities and Kathy's willingness to help.

Because we realize how difficult it is to commit time to community work, CWA is trying to develop programs with our employers which would give our workers time to volunteer for community service. We are working on a program where every employee would be given paid time off in return for volunteer work in the community. In our 1998 negotiations with AT&T and Lucent Technologies, both companies agreed to give every employee eight paid hours a month to do community service work. We are hoping that in addition to the paid time off, our members will also volunteer on their own time, and the programs will get them committed to work in their communities. Our members would make excellent mentors for young people; they could look after latchkey children, or serve as Big Brothers or Big Sisters.

Several years ago, we had a mentoring program in Alabama, developed with BellSouth. A local high school had never sent one of their students to college. Students were selected to go to work at the phone company half a day a week. Each student was assigned to a CWA mentor. Over a period of time, we saw that not only did the young person became a better student, but the mentor became a better employee. Kids began to graduate and even go on to college.

Our experience with mentoring programs shows that community service is good for the community, for the union, for the worker and for the employer. In a time when families are more fragmented and so many children are at risk, mentoring programs are more important than ever.

America has a tradition of individualism, but we also have a great heritage of community activism. When Alexis de Tocqueville visited the United States in the 1830s, he was struck by how involved Americans were in their communities. He wrote that Americans were much more community-minded than Europeans, and that really set us apart.

While things have changed somewhat since de Tocqueville made his observations, I believe that Americans still have a great enthusiasm for community. However, these days it is often difficult to get people to volunteer. With the conflicting demands of two-wage earner families, longer work hours, and other obligations outside of the workplace, volunteering is not a top priority with many people.

Our volunteer spirit may be asleep, but it is not dead. We can still inspire people to volunteer, through unions and other community-minded institutions. First you have to make sure that people understand what the need is, and how important their help can be. Most people want to make a difference, but they have to be given an opportunity.

I had a strong sense of community growing up in the Bronx tenements. We were all poor, but I never had the experience of going hungry. Still, times were tough for everyone. One childhood memory that has always remained with me was one of the few times my father struck me. I was ten or eleven years old and had just gotten a new pair of trousers. It was a Sunday afternoon and I took a ride on my friend's bicycle. I fell and tore my trousers. When I got home and my father saw my pants, he gave me a smack. I did not realize at the time what it took for him to buy me those trousers.

Life was very different then. We were not as mobile as we are today. Families were more close-knit—aunts, uncles and cousins all lived nearby. Few people had cars and they tended not to go outside of their neighborhoods. There were people I knew who had never been to New Jersey.

When I was growing up, it was a big thing to go out to the movies on a Saturday afternoon. Taking the train to Brooklyn to visit my father's family was a great adventure. On Sundays when my father was not working, which was infrequent because he usually worked seven days a week, it was a big thrill to go to Bronx Park and row on the lake.

Today, the strains on a family are so much greater, not only financially, but in many other ways. People are more mobile, much busier, and with many more distractions than before.

I feel it myself. Because of my schedule, my wife Florence is much more active in local community organizations than I am.

When we lived on Long Island, I was known in the neighborhood as "Mrs. Bahr's husband." I did not have much time to devote to activities in my own community, but she did.

Not just in America, but all over the world, families and communities are feeling the strains of rapid technological and economic changes. More than half the people in the world have never made a telephone call but you can find rows of television antennas even in the poorest, most remote villages. They see the world through television satellites and they mostly watch shows like Baywatch, which depicts the good life they think they will find in their cities. So television drives rural people to the cities all over the world, creating more economic and social unrest. When I was growing up, the only time we saw rich people was in movies, and we did not believe that anybody actually lived like that.

We will have to learn how to integrate technological and economic change so it does not destroy our communities. Computer technology is changing our sense of community, but electronic networks will not replace communities of flesh and blood. There is no question that computer technology has made distance irrelevant. Now we have the ability to converse and exchange information easily all over the world. As information technology becomes more affordable, the positive changes will outweigh changes for the worse.

However, the negative changes should be concerns to us all. We are just beginning to experience the dangers of instant information and communications. We hear about children and women being abused on the Internet, or frauds and theft being committed through advanced technology. Many of those abuses cannot be eliminated by legislation, and we all need to work harder to protect against them.

Technology can be a positive benefit in both the workplace and the community. A company that is interested in building a better community can, for example, move the work to where the people are. Instead of building a huge factory-type workplace in one community, they can establish smaller workplaces in many different communities. These workplaces are not only more convenient to workers, but also more democratic in their structure and management. With technological advancements, there is no need to have hundreds of people in the same building, being dictated to by authoritarian management. Rather, you can have small groups of people working in a more humane and more productive environment, bringing economic development to communities that otherwise would struggle.

The small workplace is better than telecommuting, as people who work at home often feel isolated from their colleagues. While various flexible telecommuting schemes may be appropriate for some jobs, we are opposed to working from home five days a week. Such isolation goes against everything that we mean by community. People want to be involved and belong; nobody likes to be a loner.

As our society changes, communities are experiencing stresses and strains. The way to make sure that technology brings us together, instead of drawing us apart, is to get people more involved with each other in the communities where they live. If people are not volunteering, experience indicates they have simply not been asked. Unfortunately, this occurs at a time when volunteer service is even more necessary than before.

As much as we object to the federal government passing more and more responsibilities to the states, it is a fact of life that we have to deal with. The transfer of the responsibility of administering social services has put even greater burdens on voluntary and charitable organizations. There are more people in need, and we must do more to help.

Community service does not have to be a major campaign. People can help one another through the church, synagogue, schools, charitable and social organizations. Community service is often very effective by the simple actions of individuals trying to make life better for others. The cleaning store in my neighborhood is owned by a young man who accepts donations of clothing for the homeless. Here is a small businessman looking out for the greater good of the community.

The Jewish faith is founded on caring for others. As a youngster you are taught to share whatever you have. A passage in the Talmud, the ancient Hebrew writings, instructs a businessman how to treat his employees. When CWA was primarily a Bell System union, there were not many Jewish members. As we broadened our membership, we began to better reflect society as a whole. I like to think that my being elected president of a union without a history of Jewish membership and being accepted as the leader of this great union all over the country indicates the tolerance of our members and leadership. I am very proud of their attitudes and the fact that they are leaders in their communities in getting across the message that it does not matter what race, religion or color you are, but who you are as a person that counts.

George Meany, former AFL-CIO president, once said: "In the final analysis, if democracy means anything in a human way, it means in the simple words of the Bible that we should help our neighbor."

Human beings have advanced more through having learned the value of cooperative caring behavior rather than individual cunning or ruthless competition.

The time has come for a return to sound values, a basic sense of fairness that must be restored to our society. These values are based on caring for others and on a genuine family concern for each other, values that define our communities as family units of Americans working together rather than isolated outposts in an economic system where success is measured in material gains without regard to others or the less fortunate among us.

The labor movement is committed to setting an example of moral leadership in our communities. We are taking the lead by showing others how to care for and to improve the communities where our families live and raise their children. Together, we can achieve our basic mission to increase the capacity of our government, our communities and our people to care for one another. One of the best ways for us to accomplish this is to become a nation of lifelong learners.

CHAPTER ELEVEN

Lifelong Learning

"President Bahr, you don't know how much you changed my life. I got my Bachelor's Degree, last month I was awarded a Masters and next month, I begin my Ph.D. And, it didn't cost me a dime."

A young African-American member, Renee Campbell Williams, an employee of AT&T, made that declaration at the tenth anniversary conference of the Alliance, our joint education and training program with AT&T.

If America is going to succeed in the 21st century, we need to become a nation of lifetime learners. A century ago, a grammar school education was enough to get a factory job. After World War II, a high school education was considered necessary for most jobs. Beginning in the 1980s, one needed a minimum of two years of college to make a decent living. Today, for the first time in our history, a majority of all new jobs require a post-secondary education.

Tomorrow's workers will be expected to have a command of reading, computing and oral and written communication skills to qualify for many of the least skilled jobs. And, all future good jobs will demand continued learning and education. Within the next few years, the vast majority of the American workforce will need the equivalent of three to five college credits a year just to stay ahead of the curve. And this trend will continue through the 21st century.

The days when an eighteen-year-old high school graduate could join a large company and work at the same job for thirty-five years until retirement are over. While many so called experts are saying that the average worker today will have seven different jobs, with different employers, I believe they are mistaken. The nature of work is changing, but not that dramatically. Instead of having seven different jobs, with many employers, the young person entering the workforce today will, indeed, probably have several different jobs, but with good corporate planning and a training program in place, it could be within the same company.

Unions can no longer promise job security. Rapidly escalating technology, keen global competition and mergers and acquisitions have dramatically changed the employment landscape. Accordingly, we changed our strategy from seeking job security to that of employment security. Employment security is achieved by providing employees with educational opportunities so they are able to improve their employability.

Education has become a key issue in our collective bargaining agreements. We consider it part of the employment security package. Those availing themselves of the opportunity for continuous learning are likely to stay with one employer. If not, they are certainly more employable.

I know how important lifelong learning is, because I am a lifelong learner.

Like so many other young men of his time, my father only had a grammar school education. My mother, on the other hand, was a high school graduate. Her father was partial to his daughters and they all graduated from high school, which was unusual at the time.

I had always wanted to get a college education. I graduated from high school and enrolled in Brooklyn College at the age of sixteen. At Brooklyn College, I made the varsity baseball team the first year, because there were no freshmen teams due to the war. A year later, I left college go to radio school to qualify myself to join the Merchant Marine.

When I left the Merchant Marine, I was married and had a child. I tried to go back to school, but the only classes I could take were during the day, while I worked at night. In the day classes, I was put in with all the recent high school graduates. At that time, teachers did not take into account that some of their students might have families to raise and full time jobs. I could not keep up with the schedule, so Florence went back to school instead of me.

As my union service progressed, and I had more contact with top level management, I began to realize that while they might have advanced degrees, they were not necessarily smarter or even better educated than their union counterparts. Some of us who did not have degrees might have felt intellectually insecure, when we should not have. We labor leaders were more than a match for management because we had street smarts.

But, as time went on and I gained more experience, I began to realize that street smarts were not enough any more. The world was becoming more and more complex, technology was advancing at a dizzying pace, the economy was becoming more globalized and more competitive. The days of the self-educated labor leader were ending. Management was smart; they had powerful information tools at their fingertips. Where once labor leaders could earn management's respect through muscle and militancy, now we also had to compete with them on the basis of brainpower and knowledge.

Union leaders and union activists need to continually educate themselves. They need to understand how the economy and the workplace continue to change, and how those changes impact their employers, our members and the union. We must be able to prepare for an unknown future, so that we can be effective advocates and leaders. Continuous education for leaders at all levels of the labor movement would help ensure that whatever changes do occur, the average worker would not be left behind.

I always wanted to go back to school, and in 1980, I finally got the chance. I was approached by Professor Harry Kelber from Empire State College in New York. Empire State is a college without walls, with centers all over the state, and flexible class schedules for working adults and other nontraditional students. Kelber had put together an educational program for labor leaders in New York City under the auspices of Empire State. The students included business agents, local presidents and staff, and others involved at some level of union leadership.

Kelber had the idea that if he could get Morty Bahr, vice president of CWA District One to enroll in his program, that my involvement would encourage others. Boy, was he right! Not only did my enrollment in Empire State encourage other union leaders to enroll, but it also showed me how important lifelong learning was. I became committed not only to continue to educate myself, but also to work hard to make sure that others were given the educational opportunities they needed to get ahead.

I entered a Bachelor of Science program. I took a battery of tests and underwent an oral interview from which the college determined how many lifetime credits I had earned from my professional experience. Within the New York State Regents system, the maximum lifetime credits one can receive is sixty-four, and I got the maximum. Still, I needed sixty-four more to graduate.

Going back to school was a tremendous experience. Everybody in the program worked hard, because they wanted to be there. All of them wanted to get the most out their efforts. The students worked for many different unions and we all learned from each other.

The class schedule was set up with flexibility in mind. The same class was given on Monday and Tuesday, so if you could not make it the first day, you could attend the next. Then a second class was given Wednesday and Thursday. And we had three hours of classes on Saturday morning.

At the same time I was going to school, I was also running the largest district in the union. I used every spare moment I had to do my homework. I remember being at an Executive Board meeting in Florida. In the afternoon, when my colleagues played golf, I stayed in and did homework.

It was a pressing schedule, but I worked hard and graduated in 1983 with my B.S. I achieved straight As in all my classes. When my four granddaughters each started college, I sent a copy of my college transcripts, with a note: If your grandpa can do it, so can you. They cried foul and maybe they were right. But I am extremely proud of all four young ladies.

I wrote my Bachelor's thesis on the 1981 election that brought 36,000 New Jersey state workers into CWA. I wrote most of the paper on an airplane while Florence and I flew to and from Tokyo. During a twenty-four hour round trip plane ride, we barely spoke.

Partly from my own college experience, but also from observing the changes in the telecommunications industry and the U.S.

economy as a whole, I saw the increasing need for CWA members to continue their educations. When I learned that I would be the next president of CWA, I had the opportunity to implement my vision to shape CWA into an education-driven union.

The Bell System had an educational program called the tuition aid plan. Participating employees had to pay tuition out of pocket and then await reimbursement from the company. All course work had to be job related. We needed a better program with different priorities. Workers should have the opportunity and receive encouragement to go back to school. The courses they take should not have to be directly related to their jobs. Since technology was advancing so quickly, the skills employees learned one year might be obsolete the next. But education, and a commitment to lifelong learning, would make them more able to adapt to almost any changes in the workplace. It would make them more productive in their present jobs and more employable if those jobs were lost.

In 1985, when I became president of CWA, divestiture had taken place and we needed to prepare for the 1986 round of bargaining in telecommunications. I started talking with Ray Williams, AT&T's vice president for labor relations. Ray was an old friend from New York Telephone, so we had a good personal relationship.

We both knew that the world was changing. Competition was escalating as the result of deregulation. Technology was advancing even faster to keep up with the new competitive forces. To stay ahead in a rapidly changing environment, our members needed to become adult learners and ultimately lifelong learners.

I talked with Ray about offering AT&T employees the opportunity to train for the new jobs that would be created in the future. In order to do so, we had to predict what those jobs were, and what skills our members would need in order to successfully fill them.

I knew we were getting somewhere when my friends in the United Auto Workers told me that Ray Williams had gone over to Ford to check out the UAW education program. Soon Ray and I came up with the Alliance for Employee Growth and Development, a nonprofit, jointly-owned corporation funded from the collective bargaining agreement, run by a board consisting of an equal number of management and union trustees and two managing directors, one each appointed by AT&T and CWA.

The Alliance was the first jointly owned, labor-management, nonprofit corporation created in the telecommunications industry. Its sole mission is to provide job training and ongoing educational assistance to workers. Because it is a 501(c)3 nonprofit, the Alliance is eligible for state and federal funding, a source of millions of dollars for employee education. Through fully prepaid tuition for job specific training or college level studies, AT&T workers qualify for other jobs in the company or learn skills for totally new careers outside the company, should that become necessary.

Since major downsizing was not foreseen in the early days of the Alliance, it was structured to prepare workers for new jobs within the company and for lifelong learning experiences. Education for self fulfillment was an integral objective.

Almost before the Alliance became operative, downsizing became a way of life at AT&T. If the Alliance had not been in place, we would have had to create it or abandon thousands of workers who would be thrown on a scrap heap.

The Alliance became expert in dealing with the problems of dislocated workers. In the early 1990s, the Department of Labor did a tracking study of laid-off workers. They found that laid-off

AT&T employees who took advantage of Alliance programs were unemployed for a shorter period of time and rehired at a much higher rate of pay than the national average.

By 1997, the Alliance was recognized as one of the most successful worker education and training programs in the nation. More than 100,000 AT&T employees had participated in one or more Alliance programs. We graduated paralegals in California, medical administrators in Virginia, and countless CWA members in other courses of study around the country.

In August 1997, we celebrated the tenth anniversary of the Alliance in Chicago, where I was surprised to be presented with an award for being the program's founder. It was during that conference that the remark was made to me that appears at the start of this chapter—how education had changed this member's life. She is but one of hundreds of members who have had education open new horizons to them.

Following the path breaking example of the Alliance, we established various forms of education programs with other employers. In 1994, NYNEX, (now part of Bell Atlantic) anticipated a surplus of thousands of employees. NYNEX CEO Ivan Seidenberg told us he did not want to lay anyone off and was willing to be innovative to accomplish it.

Through a combination of pension enhancements and Seidenberg's willingness to invest in his workforce through education, we negotiated a no-layoff agreement with outstanding education benefits.

The NYNEX education program is called Next Step. Employees attend college one day a week to earn an Associate Degree in Applied Science, with a focus on telecommunications technology. Remedial training to qualify for the program is provided to those who need it. The students work four days a week and go to school on the fifth, while getting paid for a five-day week. Next Step is no vacation. A student has to spend many additional hours a week on homework for every day that he or she spends in the classroom. The workers are excellent students—78 percent of them made the dean's list on the first cycle of the program.

The NYNEX contract also provides workers the opportunity to take an unpaid two-year leave of absence to attend college. Workers can receive up to $10,000 a year from the company in educational assistance.

The Next Step program is one of the best affirmative action programs we have. Telephone operators, mainly female, can enroll in Next Step and qualify for high skilled technical jobs for which they previously would never have had the opportunity to get.

In 1996, the Council for Adult and Experiential Learning selected a NYNEX student enrolled in the Next Step Program as student of the year. In his remarks at the awards luncheon, he spoke about how his life—and that of his family—had changed. He said this opportunity was just too good to pass up. It encouraged his wife to go back to school. And, while he, his wife and children used to spend weekends picnicking and fishing, the family now spends much of their free time in the library.

At US WEST, Pathways to the Future, a 501(c)3 nonprofit, operates similar to the Alliance. There is a major difference, however. The program is also open to management employees. I tend to believe that since supervisors have a personal stake in the future of the program, we have an extraordinarily high participation rate. In 1994, about 27 percent of the union represented employees were enrolled in college level programs.

In 1995, the Republican-controlled Congress permitted Section 127 of the IRS regulations to lapse. That made employer provided education payments taxable. Overnight, enrollment dropped almost in half. I discussed this problem with President Clinton and his staff. In 1997, Section 127 was made operable again, but Congress refused to make it permanent.

It is hypocritical to talk about the need for continuing adult education and then put financial obstacles in the path of most adults.

At Lucent Technologies (the Alliance went to Lucent when it was spun off from AT&T in 1996), 500 installation technicians across the United States are in a distance learning program in cooperation with Empire State College, working towards an Associate Degree in Telecommunications Technology. Joint programs also exist at SBC, Bell Atlantic, Ameritech, BellSouth and GTE. There is a great need to move these kinds of opportunities to all of our employers, including the public sector.

The retention rate in all of our programs is very high. Once our members return to school, they like it. They are more computer literate and more comfortable using high technology than the average industrial worker. Many CWA members have computers at home, and a majority use computers at work. I know; I hear from them via the Internet.

My involvement with continuing education led to my being selected as Chair of the Commission for a Nation of Lifelong Learners in February 1996. It was funded by a grant from the Kellogg Foundation.

The Commission, comprised of influential Americans, including Governors Evan Bayh of Indiana and Lawton Chiles of Florida, had as its mission: To change the culture of America to one of lifelong learning. Public hearings were held around the country to identify opportunities, obstacles, good practices, receive recommendations and so on.

In the Dade County, Florida hearing, the president of the South Florida Retired Executives Association testified. He told of the frustration of retired execs to be accepted as pro bono teaching assistants and as students. So, they organized and formed their 450-member association. They are now assisting teachers and are enrolled as students. He challenged the Commission to study any similar age group and predicted we would find his group has less illness of every kind. Studies are being done in Germany to determine whether continuing education also has that kind of result in the workplace.

Emphasis on worker training and education will also help insure that high-skill, high-wage jobs stay in the United States. Right now, companies are complaining that American workers are not skilled or educated enough for some high-tech jobs. Some corporate executives are pressing for legislation to lift the immigration quotas to permit them to bring more skilled workers into the United States to fill openings in high-tech jobs.

As chairman of the AFL-CIO's Department of Professional Employees, I discussed this problem with Secretary of Labor Alexis Herman. The DPE commissioned a study by the Economic Policy Institute which concluded that there was no skills shortage in this country. And a subsequent study by the U.S. General Accounting Office confirmed the EPI's findings. While there is no skills shortage at this time, there are skills mismatches. People are not where the jobs are. What we need to do is first make sure that American workers fill those job openings and then ensure that future workers have the skills necessary in a changing workplace.

Corporate America should get behind education benfits for front line workers. American employers spend approximately $30 billion a year training and educating their workers. Twenty-seven billion of that is paid by one-half of one percent of America's companies. And two-thirds of the total goes to employees who already have college degrees, in most cases management workers.

We have disproved the myth perpetuated by many in corporate America that if they train their front line workers, they will go to work for other companies with their new found skills. While some workers do move on, it is often because there are no opportunities at the companies where they work. More often than not they remain with their original employers, becoming greater assets to that company than before. They are more useful, more productive, and have better self esteem as well as more positive feelings for the company that helped give them the opportunity.

I attend many graduations of CWA members. I can tell where management has been supportive of the educational programs at the workplace. Managers show their support by attending their employees' graduations. By celebrating their graduation with them, managers prove that education is of paramount importance, and they are proud of the programs jointly negotiated between their company and the union.

The employees recognize this, too. When they get their diplomas, almost every graduate says: "I want to thank the union and the company for making this possible. I intend to go on and get a more advanced degree. I hope I can finish my career with the company, but if not, I am now in charge of my life."

I attended the graduation of 28 members employed by AT&T at Pleasanton, California. They were graduating from St. Mary's College with paralegal degrees. After the usual cracks about our country needing some twenty-eight more budding lawyers, the graduates spoke. Each said they wanted to thank the company and the union for making it all possible, they hoped they would finish their careers with AT&T, but if not, were in charge of their lives. And, finally, they were continuing for their Bachelors' Degree.

Not only were the graduates a diverse group in terms of gender and ethnicity, but they were of all ages. Thus, the families of the graduates in some instances spanned three generations—and the graduate could have been anyone, ranging from the grandchild to the parent to the grandparent.

One graduate later wrote me that this experience, including a conversation I had with her son, resulted in his deciding that school was worthwhile and not to be a dropout.

These stories can be multiplied countless times over.

Employees who are involved in learning have a better attitude toward their company. Because of their employers' commitment to their education, workers realize that they do not have to be anti-company in order to be good union members. They begin to understand that management and labor have the same goal—a profitable company. And the union will help ensure that employees have good working conditions and get a fair share of the wealth they help create. It is a win-win situation.

We need to change our basic culture to make lifelong learning a priority for everyone. Instead of thinking of school as a gas station where you stop once, get filled up and then go on, we need to think of education as the road itself.

It is not always easy. I found out the hard way that the longer you stay away from school, the more difficult it is to rebuild your learning and study skills. Perhaps even more important is to rebuild your confidence. But once adults have made the commitment, they are better students than they would have been at age eighteen. Too often younger students only learn by rote—they are in school because they have to be there, and they do not have the maturity or experience to realize the importance of education. Older students have been out in the world and know how precious an education is, and that often translates into the kind of commitment that is oriented toward results.

A commitment to lifelong learning requires a change in lifestyle and values. Instead of going out for a beer with your coworkers at the end of the shift, you might have to go to the library. Education has to become a major part of your life, almost on par with work and family. While the sacrifice can be great, the rewards are much greater. Taking advantage of educational opportunities will likely lead to a higher income, greater employment security and higher levels of job satisfaction. But the lifelong learner is also more active, better-rounded and, there is growing evidence, a healthier individual.

Lifelong learners begin to feel that their careers are more secure. They also feel more secure personally. Now they are in charge of their own lives, and they know their children are going to have better lives as well. An educated parent is going to have better educated children, in part because of the example the parent sets.

Lifelong learning should begin at birth. Parents need to be more involved in educating their children, instead of just letting the schools do it.

Committed employers must do what they can to remove the obstacles to learning. If a worker is going to school, his or her supervisor should take that into account by doing everything possible not to schedule overtime on those nights. Supervisors should be instructed to help the employee juggle the often conflicting demands of job, school and family.

Retaining a highly skilled and motivated workforce requires good management. It does not make any sense to spend money educating workers and then not taking advantage of their new skills. Just as companies have the ability to draw up short and long-term financial and sales plans, they must give equal consideration to forecasting short and long-term skill requirements and the work locations where they will be needed. Operations and human resource management must learn how to work together so that layoffs and hiring do not occur simultaneously. Good forecasting can, in most cases, avoid downsizing.

Management education needs to change, as well. In 1988, I was invited to lecture at the University of Virginia's Darden School of Business. It was a class of MBA students. I probably was the first labor leader these future management employees ever heard. I realized that a student could graduate from one of the country's top business schools and know virtually nothing about organized labor. After my lecture I met with faculty, and pointed out to them that by not including labor leaders they were turning out MBAs who were not fully rounded. The faculty saw that as a shortcoming, and every year since have invited a CWA senior staff person to attend a six-week management program at the Darden school without cost.

One year, each student was asked to invite their CEO to come and observe for a day. Gwend Johnson, the CWA staffer who was attending the program, invited me. I sat in on a class on participative management. During the lecture I was both amused and a little bit perturbed by the fact that the instructor was teaching ideas that CWA had already learned and discarded as inadequate two years earlier.

211

In order to educate people about a workplace that is changing so rapidly, we need teachers who are actively involved in those workplaces. College professors have to come down from their ivory towers and see what is going on in the real world. And we need more professionals from both labor and management going into classrooms to teach students and teachers themselves, and keep the curriculum up to date.

When teaching a subject as new and developing as workplace democracy, the teacher should have firsthand experience. You can not learn about it from a book, and in order to keep up with the latest changes, you have to be personally involved in the subject. Graduate schools, colleges, community and technical colleges, and even high schools should take more advantage of the wealth of knowledge that is available. They should be asking more guest lecturers with hands-on experience to make a contribution in their fields of expertise.

The traditional higher education system needs to change. Today, there is some resistance from university professors who feel threatened by nontraditional education programs. They should realize that the traditional university system serving young people right out of high school will never go away. In fact, it will become more important than ever as nontraditional forms of education continue to grow. By making education more democratic, a commitment to lifelong learning will actually help the traditional university by creating more customers, i.e students, for it to teach.

Many states' higher education systems are experiencing budget crunches because of lower tax revenues, higher operating costs and cuts in federal assistance. States like California and New York, whose university systems were exemplars of public higher education, are now having to cut back, and their ability to properly educate their students is compromised. But if ten million adult students, eager to learn and able to pay tuition through employer-union cooperative programs, were to enter higher education centers across the country, we would quickly be able to reverse the decline of state university systems.

Community colleges would be the first to feel the positive effects of this army of new students, particularly since President Clinton has worked to make fourteen years of guaranteed public education a goal for all students. The community colleges would then feed the four-year schools, as their students graduated with Associate Degrees and went on. When those new students earned their bachelors, the graduate departments would get a fresh infusion of talent.

By committing ourselves to lifelong learning, America will benefit from having better educated citizens who are more aware of their responsibilities and prepared to more fully participate in their communities and the economy. Educated people understand more about the world and the community in which they live. They read more books. They read more of the newspaper and understand more of what they read. They are more active in the community and local politics. They are more likely to vote and have a greater understanding of the issues involved. They understand more about their roles and responsibilities as citizens.

Studies have consistently shown that when adults participate in continuing education programs, they bring many new skills back to the workplace beyond what they learned from the course. They show a heightened ability to learn, enhanced communications skills, greater self confidence, willingness to take risks and increased capability to work in teams.

I smiled at the response of one CWA member who participated in a survey about Pathways to the Future (CWA-US WEST):

"I'm using my brain more outside of work and this is spilling over into my job."

A year after the Alliance began, Ray Williams and I asked three different local committees to give the Board a report of their activities. In each of the reports, we found that financial planning was the best attended of all classes. Ray and I asked why? The reason was so logical. Financial planning was the one class in which workers could bring their spouses, and there were no tests. It was a perfect introduction to lifelong learning for the members because it was a subject that had a direct impact on their lives, and was presented in such a way that made them very comfortable, no matter how long they had been away from formal education. Of the students who attended those financial planning classes, some 90 percent went on to other Alliance programs.

People are more comfortable among their peers and coworkers, so we try to keep them together as much as possible. But we also take advantage of technology through distance learning, where students can progress at their own speeds. Distance learning makes education as accessible and flexible as possible. Depending on the subject, classes can be held at a union office, on company premises or even in the worker's own home.

Taking a forward-looking step, US WEST now permits employees who are enrolled in Pathways programs to stay after work, use the company computer and access the Internet in order to pursue their education. This is a major step toward adults being able to get education any time and any place.

Technology can make education more accessible, provided that access to the Internet is made universal and affordable. Fourteen

western governors have gotten together to create a unique and useful program of higher education. Because the western states are
more rural and larger in area than other regions, many communities do not have convenient access to higher education. So the lawmakers established a region-wide open university where students
can take courses from different colleges in the region over the
Internet. They can take courses from any of the participating colleges, and since the schools are all accredited and the credits are
transferable, they will eventually earn a diploma recognized by the
Western Governor's Association University.

High technology and information-intensive work create an
explosion of new knowledge. Every five years the amount of available knowledge will double. But if American workers do not have
the skills to interpret that knowledge and use it effectively, then
that information will not have much value.

In 1986, when we began to develop our education programs,
there was enormous stress on math and the sciences. Hardly
anyone was talking about the liberal arts, and starting salaries
for liberal arts graduates were dropping, while salaries for math
and science majors were rising. But as technology became more
pervasive and companies began developing high-performance
workplaces, they found that they needed workers who could think
and make decisions on their own. Employees needed to be able to
work in teams and communicate with each other. Many of the
responsibilities once controlled by management were being
handed down to front line workers.

A few years ago, I spoke at an Empire State College conference in Saratoga Springs, New York that was attended by several hundred academics from all over the country. When I talked
about the renewed need for liberal arts education, there was

spontaneous applause. They know how important it is.

A liberal arts education is not just learning for its own sake, although the personal rewards are great and not to be ignored. But it is also a great asset in the workplace, as technology is advancing so rapidly that even a state-of-the-art technical education is quickly rendered obsolete. By comparison, the skills gained through a liberal arts education—communication, critical and abstract reasoning, human values, self-esteem, and most importantly, the ability to learn—will always be essential.

Because it was a monopoly, the old Bell System was able to take workers and send them to school for nine months to learn the next generation of equipment technology. No company has that luxury today. If one company does not get new technology on line as soon as possible, their competitors will. As a result, companies need employees who are able to learn to use new technology quickly.

Sometimes we have to teach the basics, like reading. The problem of adult illiteracy is too large for one union and one company to take on singlehandedly, but in one workplace CWA was able to make a big difference.

An AT&T facility was bringing in some new technology. As the union and management groups began to meet to discuss changes in the workplace, they realized that no one on the union side was taking notes. After some investigation, they discovered that even though this was one of the best paying jobs in the region, the average educational level was fourth grade, which was not sufficient to run the facility once change came. This was the Alliance's first experience with adult illiteracy, which is a problem one has to deal with very tenderly. And the Alliance did.

A GED program was introduced. I complimented the manager because he went beyond the contract giving employees two hours off if they contributed two additional hours a day.

He responded: "Don't thank me. What I have gotten back in improved productivity through the students' increased self-esteem has more than compensated."

AT&T Chairman Bob Allen and I viewed the video of the first graduation. The video was shot by the local president using a hand-held camera. It opened by saying, "the picture may be fuzzy, but the message is clear."

The class valedictorian was a woman named Sylvia, who talked about how she had worked in that plant for ten years and had never been able to read the sign in the front of the building. She told how her life had been changed by education.

As this project proved, every individual has great potential, and the best way to develop that potential is through education. Unfortunately, many people think that the chance for an education passed them by when they were young. That thinking is wrong. Our education programs are there to give everybody the chance to develop their potential. And, it is never too late.

Our toughest rivals in Japan and the European Community develop the talents of all their workers. If we are going to effectively compete against them, the United States also must bring out the best in all of our people.

In the years ahead, I believe we will see expanded activity by the entire labor movement in worker training and education. I recommended to Lane Kirkland, when he was president of the AFL-CIO, that the labor movement needed a Masters program that would deal with subjects such as technology, the global economy, and their impact on our members and unions. Lane appointed me chair of a special committee and in 1996, we established the first program for an MA in labor studies at the University of Massachusetts. I was proud that Ralph Maly, assistant to CWA Vice President Irvine, was one of the first students to enroll.

The work of the AFL-CIO Education Committee traditionally focussed on public education, with an emphasis on elementary and secondary schools. But in recognition of the growing need for lifelong learning, President Sweeney split the Committee. One Committee will deal with the role organized labor plays in helping to improve public education. The other, the one I chair, will deal with the development of skills standards, worker training, education and lifelong learning.

The Degree Program at the George Meany Labor Center usually has more students from CWA enrolled each semester than from any other union. In addition, the George Meany Centers, both East and West, are training and educating union leaders in a large variety of union related subjects.

CWA also sponsors several education programs. The Joe Beirne Foundation has been awarding sixty scholarships a year to CWA members and their children. In 1998, we were able to use royalties from the AFL-CIO Union Privilege Benefit Fund to expand the number of scholarships to one hundred thirty and to expand the program to include members' grandchildren.

It is often difficult for full-time staff to take on the additional responsibilities of class work. We try to remove the obstacles. We prepay all tuition expenses. We give them excused time when they need to be on campus so they do not have to use vacation time. And we will give a vice president assistance to cover for the staff representative when in school.

Ideally, I would like to see education programs, including the financing, taken out of the collective bargaining environment. Education programs must become institutionalized and not something a company could stop, even if workers went on strike.

When the State University of New York in 1996 honored me with a Doctorate of Humane Letters, it was one of the proudest moments of my life. Recently, I was standing in the hallway to my office with Dr. Michael Maccoby, a friend and pioneer in workplace democracy. He pointed to that honorary degree, which was hanging on the wall along with photos of me with President Clinton and Vice President Gore. "This is what you should be most proud of," Dr. Maccoby said.

"It is," I told him.

When I gave the commencement address at SUNY, I spoke of the need for lifetime learning. I told the graduates, which included students earning associate, bachelor and masters degrees, that their education was just beginning.

Afterwards, at the reception a group of graduates sought me out. One said: "We enjoyed your speech, but we're sorry you are here."

"How come?" I asked.

"We thought we were through with our education."

I told them that this was just one stop on a train that keeps going—a lifetime of learning.

CHAPTER TWELVE

Democracy in the Workplace

Y ou've probably seen the episode of *I Love Lucy* where Lucy is working alone on the assembly line of a bakery. The cakes keep coming down the line, faster and faster, as Lucy struggles to put icing on them and pack them into boxes. She can't keep up, and the harder she tries, the more mistakes she makes and the farther she falls behind. Finally Lucy just gives up. She stands there crying while the cakes pile up on top of each other. It's a funny scene, but for many American workers, the humor stings, because Lucy's bakery is a lot like their workplace.

For more than a century, the American economy was based on industrial mass production—employees performing repetitive work along an assembly line, overseen by authoritarian managers. In his 1911 book, *Principles of Scientific Management*, Frederick Winslow Taylor established a rationale that served as the basis for industrial management. Taylor distinguished between "thinkers" and "doers." Thinkers were management. Doers were workers who were basically told to check their brains at the factory gate.

Under Taylorism, work was broken down into its smallest components, and reduced to a series of repetitive tasks. Workers were told what to do and when to do it, but never why. As Taylor himself once told a factory worker, "You are not supposed to think. There are people paid for thinking around here."

The Bell System used Taylorism with a vengeance. Elton Mayo's famous Hawthorne studies of the 1920s were performed at a Western Electric factory in Hawthorne, Illinois. The Hawthorne project measured working conditions down to such details as what color to paint the walls to make people work harder, and Mayo's conclusions were largely implemented throughout the Bell System.

While Taylorism was the dominant management theory of the 20th century, technological advances and the global economy have made it an anachronism. Today's global economy demands high-value, high-quality products and services. These objectives can only be achieved in high-performance workplaces that promote innovative practices, employee involvement, and power sharing at the ground level.

We call this democracy in the workplace. As more corporate managements recognize and accept that Taylorism is dead, I believe the future of labor-management relations will be built on greater employee decision-making on the job and labor-management power sharing. By empowering front line workers to make their own decisions about their work, we will make American workplaces more humane and more productive at the same time.

Workplace democracy means allowing workers to make decisions they are best qualified to make. In order to make those decisions, workers must have information and knowledge about not only their specific jobs, but the entire work process and the employer for whom they work. Workers and the unions that represent them need to have a significant voice in both the day-to-day management of work and the long-term strategies of their company.

To our union, workplace democracy is a natural extension of the democratic principles that serve as the foundation of the collective bargaining process. Worker participation cannot be used as a veiled attempt by management to undermine collective bargaining agreements or our roles as union representatives.

A union workforce is absolutely essential to making democracy in the workplace a success. As the organized voice of the workers, unions provide the solidarity, the unity and the structure to help reorganize the American workplace. Unions legitimize employee involvement programs in the eyes of the workers. Union contracts provide the necessary security to encourage workers to speak out freely and to criticize and change existing practices. Unions provide a mechanism through which the experience, knowledge and ideas of employees can be channeled to management in a coherent fashion.

Unions also make sure that productivity gains and profit expectations are appropriately balanced with job security and quality of life goals. It should not be a surprise that most workplace reorganization programs that were unilaterally developed and administered by management have failed over the years.

A union brings stability and a long-term perspective to workplace reorganization. To us, power sharing and employee involvement is no fad. We want to insure that workplace democracy lives beyond a particular corporate executive and survives changes in leadership on both sides.

Democracy in the workplace represents change that can be very threatening. Some managers view any kind of power sharing arrangement as an erosion of their management rights or a threat to their authority. To a union, cooperative relationships have the potential to undermine our role as representatives of the workers. In other industries, management has tried to use quality circles, total quality management and other employee involvement schemes to side-step the union and erode its strength.

But strong unions have nothing to fear from cooperation with management. In fact, there are times when we can achieve important gains for our members and our union that would not happen through an adversarial relationship.

Of course, no matter how much labor and management cooperate, there will always be areas of disagreement. Unions represent the interests of workers and management represents owners. While we do have many interests in common, we have values that will inevitably conflict. Our objective is to increase the area of common interest and make things work better for everybody.

In developing workplace democracy programs, flexibility and cooperation are vital. Employee empowerment is an ongoing strategy in which labor and management must learn from experience. Workplace democracy requires enormous trust and understanding between labor and management. Building and maintaining that trust requires great effort and enormous patience.

Greater worker participation for CWA members was one of the goals I set for myself when I became president of the union. But the first steps toward workplace democracy came well before I assumed that office.

In the late 1960s, our members in the Bell System experienced a myriad of problems in workplaces around the country. We called them "job pressures." President Beirne invited top AT&T labor relations management to meet with the CWA Executive Board in 1972 to discuss these job pressures in more detail. After a year of investigation, management confirmed the existence of virtually all of these problems, but said they were unable to effectively deal with them since they were all different and were occurring at thousands of workplaces around the country.

June 15, 1979 was declared "Job Pressures Day" in a nationwide effort to demonstrate our concerns. Informational picketing took place all across the country. While this activity did not bring any immediate relief, we did lay the groundwork for a breakthrough in the 1980 round of national bargaining with the Bell System.

We negotiated ten guidelines designed to develop joint labor-management committees that would deal with problems at individual workplaces. These guidelines later became known as Quality of Work Life, or QWL. At the time, QWL was considered very innovative and received great attention from the news media. The QWL process brought union leaders together at the corporate and line levels as we began to learn how to work more cooperatively together.

Dr. Michael Maccoby, one of the leading consultants on workplace democracy, was instrumental in helping us develop QWL. Glenn Watts initially retained him to perform a study and eventually develop our 1980 contract language in this area. Since then Dr. Maccoby has worked closely with CWA and AT&T in developing our workplace democracy programs.

When I first met Dr. Maccoby, I must admit I was a bit skeptical about what they were then calling participative management. It sounded to me like they wanted the union to become shills for the company. I came out of the traditional labor movement, having had to fight to get my workplace organized in 1954 and continuing to fight ever since. I was not convinced that we needed to work more closely with management; I thought we needed to be fighting harder.

I told Dr. Maccoby that on a scale of one to ten, I was probably a two in enthusiasm about this project (actually, I was about a 0, but I did not want to discourage him). Still, I wanted to see how my local leaders responded to what Dr. Maccoby had to say, so in 1981 I invited him to my regional meeting.

I was most candid at that meeting. I told the local leaders that I was not sure that this was the way the labor movement should go in the future. But since I was a member of the committee that negotiated it, I felt morally bound to give it a try. And, I asked them to do so as well.

Maccoby started generating some interest and, working closely with Larry Mancino, then New York director, they began developing examples of union-management participation. I told them both that if they could show me that the program would really help workers, then I would support it.

Once I began to see that, rather than just being an abstract theory, worker participation could actually bring tangible benefits to our members, I began to support Maccoby's efforts. Dr. Maccoby and I visited an operator center in Grand Rapids, Michigan. I saw how much the operators, who had been involved in a participation program, were excited about having a say in the way their workplace was run. On the way back, I told Dr. Maccoby, "Now I'm becoming a believer."

By 1984, about 100,000 employees throughout the Bell System had participated in QWL training or work teams. The QWL process was virtually disbanded with the court-ordered breakup of the Bell System in 1984. And during its brief history, there were shortcomings in QWL. In many locations, the process never really developed beyond the surface concerns of workers, such as what color to paint the walls. Even as the process evolved, management still retained control and authority over work processes with little input from front line workers. While employees were allowed some input, real power sharing did not take place. The success stories were few—but were there.

We also discovered that we had not done a very good job of instilling union values into the process. Some of our members who became QWL coordinators began to identify more with the company's goals and less with their coworkers. Indeed, when the 1983 strike occurred, we found some QWL coordinators, CWA members, crossing our picket lines because they identified so closely with management.

The Common Interest Forum, negotiated in 1983, outlasted QWL and proved to be even more significant. The Forum was comprised of high-level union officials and corporate executives who met periodically to discuss long-range issues impacting the company.

The Forum was the setting where CWA first began to talk about the need for creating a company-funded, jointly owned training corporation to prepare workers for new job skills in the company and for continuous learning. This vision was realized in our 1986 contract with AT&T that created the Alliance for Employee Growth and Development.

In part as a result of the Forum's discussions, in 1989 bargaining we negotiated a package of family care benefits that set the standard for the private sector and is still unmatched in any other industry. At AT&T, for example, we won a multi-million dollar developmental fund which provides grants to initiate and expand local child and elder care services. These benefits have now been negotiated into our contracts with other employers.

We also agreed to build a managed care network with AT&T

to deal with the rapid rise of health care costs. This was the first and is still the largest single private sector managed health care program in the nation. We have negotiated similar managed care programs on a regional basis with the operating companies.

As we approached 1992 bargaining, we realized that the time had come for a new vision for unions in the telecommunications industry, a role that moved us away from conflict and toward real participation and power sharing. We took the first steps toward realizing that vision in our new contract with AT&T. The contract set forth a new model of workplace democracy called Workplace of the Future (WPOF), which, for the first time, guaranteed workers a significant voice and a role in managing change in the workplace. I was now a believer.

Workplace of the Future was launched at a joint CWA/AT&T conference on March 8, 1993, involving nearly 1,000 local union officials and company managers. It was the largest joint labor-management conference that we have ever held. Top corporate management and top union leadership spoke about our commitment to the process in a display of support at the highest levels of leadership in both our organizations. This was critically important to getting off on the right foot. Secretary of Labor Robert Reich was the keynote speaker and he was highly supportive.

The key part of WPOF is that it is voluntary. Management and the union at all levels must mutually agree to participate in WPOF. Either party can walk away from it at any time. In accordance with labor law, the union has the power to select all WPOF teams and committees. This right also is specifically written into our collective bargaining agreement. The union has a visible and active presence at every level of WPOF so there is no doubt that CWA is an equal partner in the process.

Although WPOF is a revolutionary concept, we implemented it in an evolutionary fashion. In launching WPOF, our initial efforts were concentrated on communications and education. We had to communicate top corporate and union support for WPOF and educate local union leaders and managers on the principles that serve as the framework for WPOF.

The process works through the creation of business unit planning councils and workplace teams comprised of equal numbers of managers and union members. Business unit planning councils jointly decide where the business should go in the future and how best to meet their customers' needs. Workplace teams develop strategies to best achieve these goals. CWA participants make these decisions on an equal basis with management. No changes are permitted in contract language unless it has been approved by the Constructive Relationship Council, which includes the CWA vice president in charge of the AT&T bargaining unit and his or her counterpart within the company.

At the highest level of the process is the Human Resources Board, which was created to include top union and company executives and two outside experts in the field of human resources. The HR board meets periodically to take a long-term look at business, broad strategies and global human resources and business issues.

WPOF has survived many tumultuous events in AT&T— layoffs, downsizing, increased business pressures, the trivestiture and changes in top executive leadership. I am glad that so many managers and local union leaders remain so committed to the process.

After WPOF was established, Dr. Maccoby invited me several times to the Kennedy School of Government at Harvard University, where he ran a program on technology, public policy, and human development. When asked to speak, I would tell the story of how my support for workplace democracy was tepid at first. "In terms of enthusiasm, I started out a 2," I would say, "but now I'm a ten."

How does AT&T feel about Workplace of the Future? This story best illustrates management's attitude. Top GTE management approached me about developing a new relationship with CWA based on what we are doing in WPOF. GTE asked Dr. Maccoby to work with the company and the union on this project. He went to AT&T top management to make sure they did not see a conflict of interest. The response he got was this: The company had no problem with him helping to get the process started. But they asked him not to continue beyond the start-up phase.

AT&T top management sees a competitive advantage in their strategic partnership with CWA, a significant value-added component of their corporate strategy.

We certainly agree with AT&T. Workplace democracy programs required high levels of mutual trust that recognized shared interests and common goals between labor and management. In addition, successful programs reaped positive benefits for all of the company's stakeholders: workers, management, customers, and investors.

After a series of layoffs and shakeups, morale at AT&T was lower than a snake's belly, according to the company's own 1997 survey. Our members felt overworked, overburdened, overstressed and overpressured. These results were not surprising to us.

What the survey also revealed, however, is the power of joint labor-management participation to overcome these deep seated negative employee attitudes. One of the highest rated company activities is the availability of job training and education opportunities. For our members, these services are provided through the Alliance.

Another aspect of the survey worth noting is the fact that those employees who participate in WPOF scored higher in fourteen out of fifteen satisfaction categories compared to those who did not. This finding is a powerful argument to overcome the reservations of even the most cynical manager or apprehensive union representative about Workplace of the Future.

The union has used WPOF to save jobs, enhance the quality of jobs for our members, save shareholders millions of dollars and improve customer service. WPOF has also been the forum through which CWA communicated our intense commitment to management that union members wanted access to all jobs within the company, including jobs that had been "designated" nonunion. Our progress in finally achieving union wall-to-wall language in our 1998 contract with AT&T began through the WPOF process.

Workplace of the Future has been recognized by many scholars as the most advanced worker participation program in which the union plays a major role. That distinction is crucial. Every piece of empirical information now available demonstrates that workplace democracy functions most effectively where the union is an equal partner with management in the process. Unless the union is a leader in the program, true democracy is not possible.

Education is a vital component to any successful workplace democracy program. AT&T Labor Relations Vice President Bill Ketchum and I visited the AT&T operators service center in Richmond, Virginia to recognize the accomplishments of half a dozen operators who were graduating from Alliance-sponsored educational programs. Once we arrived there, I realized that something very exciting was going on. Everybody's attitude was very upbeat, and I could not tell if the woman chairing the meeting was union or management. I told them that I had a sense that good things were happening. One union member said, "That's because we run this facility. This is a self-managed operator center."

The union member pointed to another woman. "She's the manager, but she's really our coach. We don't have any other managers here. And you know what? We have the best customer service satisfaction rate in the state of Virginia."

In the Richmond facility, the employees ran the office. They determined, for example, whether overtime was needed, and then they assigned the overtime. Having been given the responsibility to run their own show, these workers responded with performance and pride, and the union played a major role in restructuring their workplace. So, it is no surprise that AT&T's operator services unit, 100 percent union, won the U.S. Commerce Department's Malcolm Baldrige Award for Quality in 1996.

I visited an AT&T (now Lucent Technologies) factory in Dallas where the entire plant had been remodeled. If you saw the way the place looked before (and I did, having a photo of the old factory) and compared it to the way it looked now, you would have thought the remodeling had taken six months during which time the plant must have been closed. Instead, they were able to completely retool and remodel without losing a single day of production.

When I walked the floor and asked workers what they did, they not only told me their job, but also told me what the next person's job was, all the way to the final product being sold to the customer and that customer's name.

In 1995, the Dallas factory became the first American factory ever to win the Deming Award for Quality.

Workplace democracy is not confined to the private sector. Instead of turning to privatization and other forms of contracting, government leaders should be exploring new forms of workplace democracy with their unions.

Unions bring value to any workplace, including government. In 1995, I spoke at a conference at Rutgers University arranged by Governor Christine Whitman. Among other purposes, the conference explored ways that CWA and other unions could engage in cooperative efforts with the state.

Even prior to this conference, CWA public sector locals had engaged in several cooperative programs that benefitted state workers and taxpayers in New Jersey. At the Department of Environmental Protection, for example, CWA Local 1034 worked with management for more than a year to develop an alternate work week program. Working together, labor and management came up with a program that allows employees to work either four days a week, or a nine-day period. The program was so successful that plans were considered to expand it to other agencies.

Local 1034 also worked with management to modify the voluntary furlough program. Workers could take up to ninety furlough days a year for work, family and education needs. If a female worker, for example, had a baby, she could use the voluntary furlough program to work three days off and two days on. This way, workers could avoid going on a part-time status and thus losing benefits.

We established a positive track record of cooperation where management worked with us. I wish I was able to say that my remarks led to a renaissance in labor-management relations in New Jersey. Unfortunately, they did not. But we are not giving up. It remains in the interest of CWA members and the interest of government employers to establish a positive working relationship that will benefit all of us.

Workplace democracy also serves as an organizing tool. Since we began Workplace of the Future, nonunion AT&T workers in right-to-work states have joined the union so they can take part in the program. Where CWA engages in worker participation programs, we have won agreement from our employers that union representatives will have appropriate access to unorganized workers, and that these workers will have a free and fair choice of whether they want CWA as their collective bargaining representative.

We hope that workplace democracy becomes a primary concern for the entire labor movement. According to the Dunlop Commission, nearly two-thirds of American workers want more say in workplace decisions, and it is up to labor unions to provide them with a voice. I serve as chairman of the AFL-CIO's Committee on Workplace Democracy, whose purpose is to encourage unions and their members to work with employers to give workers more decision-making power on the job.

The AFL-CIO recently challenged business and government leaders to join with the labor movement to redesign the American workplace, to change the way work is organized and develop a different relationship between management and labor. The 38-page report on *The New American Workplace: A New American Perspective* showed that the AFL-CIO is committed to workplace democracy and really focussed the labor movement's attention on this crucial issue. Now progressive labor leaders are no longer debating whether to develop worker participation programs, but rather which programs they should emulate.

While organized labor honestly believes that in order to be

successful in the next century, we must have a better cooperative relationship with employers, many in corporate America have not caught up to us yet. In order for workplace democracy to succeed, we need support from the very top. As a union, we need to be able to tell our members on the shop floor that the CEO of their company has totally bought into the program and is willing to work with us, sharing power and information in order to reorganize the workplace.

"I'm not committed to this new working relationship because it's some comfortable, sentimental, egalitarian vision," Stan Kabala, AT&T vice president, said at Workplace of the Future's inaugural conference. "I'm committed because it's the best way to satisfy our customers and grow the business."

We need support like this for worker participation to be more than just another corporate cliche.

Once they embark on a participative relationship, management finds out very quickly that the people doing the job know a lot more about it than anyone else. And when they begin to involve that worker, there is more interest and greater pride in the job.

Workplace democracy empowers workers on the job, in the union, and in the community. It gets them more involved and mobilized. Because the union is actively involved in the administration of the programs, workers have more contact with and involvement in their unions. They are no longer passive observers. Now they are active union members.

This is a natural extension of the traditional role of the labor movement. Since union workers enjoy the right to participate in decisions affecting their jobs, they also have a responsibility to take a leadership role in participating in their community.

Workplace democracy is a process, not a plan. There is no one single model that works in each situation. We are seeking long-term, permanent changes in our culture and our ways of doing business with each other. That will not happen overnight.

Because it is a process, no one can say for certain where workplace democracy will eventually lead. One thing we do know is that it will change the entire power structure of large workplaces. We will see more worker involvement and higher productivity through flexibility, constant learning, communication and improvement.

Workplace democracy programs will also change collective bargaining. Through these agreements we have built the foundation for a living contract that does not freeze us into a static system for a three-year period, but which offers ongoing flexibility and responsiveness for both the union and management. Some significant form of profit-sharing must eventually develop out of worker participation. When workers contribute to greater efficiency and productivity, they should share in the savings through higher pay raises, greater job security and improved benefits.

I see a day at AT&T, SBC, GTE and others when we hold meetings involving management and union representatives from all of the countries where those companies do business around the world, not to engage in collective bargaining, but to share information, discuss their global objectives and the role of the unions in achieving those objectives at each of the companies.

Workplace democracy offers CWA and its employers the opportunity to set an example for the rest of the telecommunications industry and perhaps for the entire private sector. In the global marketplace of tomorrow the successful company will be known for the quality of employee that it keeps rather than the numbers of workers who are laid-off.

As service becomes more important than ever, management needs to rely more on the expertise of those closest to the customer—their front line employees. Workers and their unions are the greatest resource that a company has. The companies that cultivate this resource are the ones that will prosper. At the telecommunications companies, for example, customer service representatives and technicians have direct contact with customers and generally are more aware of their concerns than management. These employees have first-hand knowledge of their customers' needs. Their employers should be listening to them.

It is no surprise that companies that draw upon the enormous resource of their employees' knowledge and experience are the ones who have been most productive and most profitable. The success of CWA's members in making their companies more competitive debunks the myth that union companies cannot compete in the global economy and that a union workforce is a net cost rather than a net benefit. In fact, the very opposite is true.

"CWA has been very helpful in making Ameritech competitive," says Ameritech CEO Richard Notebaert. "We are beating the nonunion shops and winning in the marketplace. The union and management have created flexible situations that have significantly helped to improve Ameritech's profitability."

There is no question that the union can bring value to the company where management is willing to work with us. In 1989, the National Women's Political Caucus honored CWA and AT&T for our major breakthrough on family issues in our contract that year.

At the awards ceremony, AT&T Vice President for Labor Relations Ray Williams said, "We've been prompted by the union for a long time to do these things, but what finally pushed us over the hill to do it was the competitive marketplace. We began to realize that all the things the union was telling us were true. We don't want an employee who is worried about a sick child at home dealing with a $50 million customer."

I sent a copy of his speech to Lynn Martin, then-Secretary of Labor, and told her that if business people do not want to provide decent family benefits because it is the right thing to do, tell them to do it because it is good business.

Other CWA employers realize the competitive edge that we can offer them in the marketplace. In his speech to the 1997 CWA convention, SBC CEO Ed Whitacre spoke about how the union assisted his company in the various states and that he might not have been successful in purchasing 30 percent of South African Telecom without my personal assistance.

But popular myths about unions and workers die hard. Some still believe that unions are inflexible, tradition-bound and unable to respond to changing economic conditions. There is a prevailing belief among corporate leaders and the public that unions hurt productivity and raise labor costs. They think that our economy would be much more competitive if there were no labor unions.

They could not be more wrong.

A union workforce brings value to all the stakeholders of a company—employees, management, executives, shareholders, community. The union helps working families by raising their wages and benefits and improving their working conditions, whether they belong to a union or not. The union helps the customer by improving the quality of service or goods produced. The union helps the economy by ensuring that workers earn a decent living and are able to contribute as both producers and consumers. The union helps the employer by providing a stable and flexible workforce focussed on long-term goals of productivity and profit. And finally, the union helps society by maintaining living standards and being involved in community service.

CWA and every other union represent working people. We want to see them succeed. If we didn't add value to a company, our members would not have jobs for long. Labor and management have the same goals: a profitable company that provides secure, good paying jobs today and creates job opportunities for the next generation. When our employers are productive and profitable, that translates to more and better jobs for our members. So it is in our best interests that our employers prosper and grow. And it is in their best interests to recognize the value a union workforce brings, and help the union grow along with them.

Looking at different ways of doing business together will take time. Labor relations in the U.S. have been marked for so long by adversarial relations that some union leaders and managers know of no other way to relate to one another. This needs to change. The real challenge in labor relations is how to solve problems and come to a fair agreement. That takes commitment, creative thinking and shared respect and honesty between labor and management.

As labor relations change, so must the way our workplaces are organized. To compete in today's world, management must develop new ways of organizing work that frees workers from the yoke of oppressive managers who believe that all power and decision-making must be concentrated in their hands.

As we have learned from our toughest foreign competitors, new forms of successful work organization will require workers to cooperate more with each other, bring their brainpower to work and permit workers direct responsibility over how they do their work. To do this, workers must be empowered to make their own decisions on the job, including the power to make decisions that were once reserved solely for management.

Workplace democracy unleashes the collective expertise and creative talents of union workers. Unions will add even greater value to our employers by helping to create productive, profitable and satisfying workplaces for our members. There is no better formula for higher profits, sales, service and customer satisfaction. And workplace democracy will be increasingly important as we move to organize the new American workforce of professional and technical workers.

CHAPTER THIRTEEN

The New American Workforce

I watched the Screen Actors' Guild Awards and was very moved to witness the industry's top performers attend their union awards dinner. These professionals make millions of dollars a year, and yet they were openly proud of their union. They had not forgotten where they came from.

During the ceremony, Gloria Stewart, who was nominated for a Best Supporting Actress for her role in *Titanic*, spoke about what working conditions were like for actors before she helped form the Guild.

Jack Nicholson told about the ten years that he worked, when he worked, earning only the Guild minimum—which is what most of them still earn today.

Most Americans don't think that performers or other artists need union protection. We only hear of the millions a singer or some other performer earns for their work. But before the union, these people were among the most exploited workers in America.

The so-called "Golden Years of Hollywood" were not so very golden for many actors and actresses who were forced to virtually sign away their lives to the movie studios. Even today, only a few attain the stardom of a Jack Nicholson. For many, their union is the only thing that stands between them and being at the complete mercy of producers or studio owners.

I had a chance to meet with our ambassador to the Czech Republic, Shirley Temple Black, when I was in Prague. She was proud of her membership in the Screen Actors' Guild and recounted that had it not been for the union, she would not have had the insurance she needed to take care of her mastectomy since the money she had earned as a child star was gone. She spoke about the number of jobs she lost because she wouldn't cross a picket line. I couldn't help but wonder whether President Reagan, who had appointed her, was aware of her union loyalties.

The late great newspaper columnist Heywood Broun spent a lifetime championing the cause of ordinary people. But he had only to look around the newsrooms of America in the early 20th century to see that the members of his own profession were among the most vulnerable workers in America.

Joined by other concerned journalists, Broun was one of the founders of the American Newspaper Guild, which later became The Newspaper Guild and is now affiliated with CWA. Writers, editors and other newspaper workers quickly came to value the protections of a union contract.

Broun was among the highest-paid journalists of his era. But he never lost sight of who he was or the heights that his profession could reach. He was not afraid to stand up for those less fortunate, regardless of the risk to himself.

"After years of holding down the easiest job in the world, I hated to see other newspaper men working too hard. It embarrassed me more to think of newspaper men who are not working at all."

Can you imagine one of today's media superstars speaking or writing those words? CWA is deeply honored to be entrusted as the keepers of the memory and the spirit of Heywood Broun.

Our TNG members are among the actors, broadcast technicians, computer software specialists, teachers, nurses, and other health care specialists—millions of American workers who are transforming our economy. They are a new American workforce and they are in need of union protection just as much as any other worker.

The change in how Americans earn a living has been nothing short of astounding over the past half-century. In 1950, the goods-producing sector of the economy employed some 20 million Americans while nearly 30 million worked in the service-producing sector.

Today, that spread has widened astronomically. Nearly 100 million Americans are employed in the service sector while manufacturing has remained stagnant at about 20 million. The workers we think of as white collar comprise more than 58 percent of the total workforce. Traditional blue collar work amounts to barely 28 percent of all jobs.

This trend will accelerate in the years ahead, as the nation makes the transition to a knowledge-based economy. Over the next ten years, professional jobs are expected to grow by 26.6 percent; technicians and related support occupations by 20.4 percent; marketing and sales by 15.5 percent. Production-type work will increase by barely 7 percent.

Obviously, if organized labor is to grow, we must look to the new workers in the professional, technical and administrative support jobs; the "new collar" workers who will dominate our economy.

Labor has a moral obligation to bring the benefits of collective bargaining to agricultural workers, janitors and other low-wage workers. And we should. But I also believe that future historians will judge the success of labor's resurgence at the dawn of the 21st century on our ability to attract greater numbers of professionals and technical workers into our ranks.

One of the challenges facing the labor movement will be restructuring our organizations to meet the needs and desires of the new American workforce.

Contrary to the myth that professionals don't join unions, these workers can be organized. "White collar" workers now make up 45.7 percent of all union members. Most of them, to be sure, are employed in the public sector. But the fact remains that teachers, engineers and other highly trained, educated workers are turning to union representation in record numbers when given the opportunity.

The reason is simple: there is strength in union organization.

Gene Upshaw, president of the National Football League Players Association, explained it best to me. CWA had a strike at a hospital in Buffalo which was supported by the local Jobs with Justice coalition. We had a rally for the hospital workers who are part of this new American workforce that labor must successfully organize.

The hospital was not about to roll over for the union and the hospital workers were fighting hard for their futures, putting their careers and their livelihoods on the line. Lynn Williams of the Steelworkers and Gene Upshaw joined me at the rally.

Gene had not yet been elected to the Football Hall of Fame, but he followed up his great career with the Oakland Raiders by becoming involved in the NFLPA. Now, surely there were other places where he could have been that Saturday morning instead of speaking at a rally in Buffalo for a group of striking hospital workers.

But it was important for him to be there, he said. He was supporting the CWA members because he knew that we were all in it together, professional athletes, nurses, orderlies, every working person. He told stories about what it was like to play football before they had a union. (And this was also before the football players had won free agency, so most of them still earned relatively modest salaries.) They had no pension plan. Their careers were short and could be ended by injury at any time. When a player's career ended, they often had nothing to fall back on. If a player got injured, he did not get paid. The working conditions also were horrendous. Gene told how Oakland played one game at Green Bay in sub-zero weather and the coaches wouldn't let them wear gloves, in order to show the other team how tough they were.

The union changed all of that. Fans often forget what conditions were like for all professional athletes before their unions grew strong enough to challenge the owners. Obviously, most owners care little about the players' unions. Peter Angelos of the Baltimore Orioles is a rare exception.

In 1994, when the baseball players struck to protect their right to free agency, the owners threatened to field replacement players. Angelos was the one holdout and pledged to close the Orioles stadium rather than allow scabs to play there. Needless to say, Baltimore also is a strong union town. And labor support helped the players in the communities where union membership is high.

As Gene Upshaw said, we are all in this together.

Upshaw is not the only successful professional to become a successful union leader. The great Broadway and movie actor/singer Theodore Bikel is now head of Associated Actors and Artistes of America, which represents the performing unions. He manages to balance his career with his union service. When I first met SAG President Richard Mazur, who also is a well-known movie and television actor, he said to me, "Gee, I envy the work you do." I responded that it was I who envied the work that he did, both on screen and with his union.

In recognition of the importance of professional workers to the labor movement, former American Federation of Teachers President Albert Shanker led the formation of the Department for Professional Employees (DPE) more than twenty years ago. The DPE is an autonomous, constitutional department of the AFL-CIO. It serves as the coordinating arm for unions representing professional, technical, entertainment and administrative support workers.

In 1997, I was honored to be elected chairman of the DPE, which reflects the enormous shift in the membership of CWA. My colleague, Jack Golodner, serves as DPE president. As AFL-CIO President John Sweeney said during a DPE meeting: "The labor movement must relate to the concerns of the new majority of workers, embrace their causes and vigorously recruit them into the ranks of organized labor."

The unions of the DPE are committed to do just that. But it is clear that new approaches are necessary to attract these workers. They have concerns that are not in the mainstream of daily labor activity. Writing in *Working USA*, David Moberg recently noted that, "For many workers in the bottom rungs, the old standbys of wages, benefits and a modicum of job security are critical. But the more technical or professional the work (as with nurses, doctors and university personnel who are showing increased interest in unionization), the issues of career and the conduct of work gain more importance."

Professional and technical workers, however, are not immune to the rapid technological advances that have affected other workers. The popular image of professional and technical jobs is work that is secure, based on skill and years of formal training or education. These are occupations where the individual had a great deal of personal control over the work, with wages that went up. Technology has made these jobs increasingly rare in today's world of work.

Technological advances that had long affected production workers now have a significant impact on professional and technical occupations. This is a real change from the past when professional and technical workers often benefitted from technological change. They were able to use technology as an extension of their brain power to enhance their personal skills, productivity and creativity. They had considerable individual control over their work.

Where engineers and highly skilled technicians enjoyed interfacing with workers in a factory, today the computer has them confined in a cubicle with virtually no personal contact with production workers. Just as computer-driven technology deskilled work for blue collar workers in the 1980s, it has now impacted the professional and technical employees.

We have seen the rise in a temporary and contingent workforce employing hundreds of thousands of professional and technical workers—none of whom accrue seniority, vacations, pensions or sick benefits. Companies can easily opt to send their programming work to low-wage offshore employers.

Wages remain stagnant or flat. According to the Economic Policy Institute, wages for new college hires in business administration dropped 4 percent from 1989 to 1996; 8.5 percent in civil engineering; 9.2 percent for computer programmers; 14.5 percent for biology majors.

Career ladders are being destroyed as professionals at all levels are hit by these massive changes. Job insecurity is sweeping the professional and technical occupations as never before and they are looking for alternative solutions to their problems.

Professional and technical occupations constitute the fastest growing segment of the U.S. workforce. In the short term, professional specialty occupations will account for one-fourth of all new jobs created by 2006. If we are to be successful in bringing the benefits of trade unions to this work group, it is essential for us to understand their special needs and perspectives.

The DEP commissioned Professor Richard Hurd of Cornell University to conduct a survey of the attitudes of this growing new work group. The findings are useful.

The 1,500 employees in the study work for a variety of employers and in many different settings. They report a very high level of job satisfaction (83 percent). The reason for satisfaction most mentioned is the type of work they perform. Attachment to their occupation is also very high, with ten years of experience or more for 73 percent and 74 percent expecting to be in the same occupation five years from now.

It is noteworthy, however, that this commitment to their work does not imply approval of employer actions. Top management is given 56 percent negative ratings.

It was no surprise to learn that the most important work-related issue to these workers is freedom to exercise professional judgment.

For professional and technical workers, the key attraction of an employee organization is that they give workers a voice. The key for not joining is concern that this may create conflict at work. Eighty-one percent of the workers surveyed believe that employee organizations should seek to develop a cooperative relationship with the employer.

The challenge to CWA and other labor unions is that among employee organizations 36 percent of professional workers, according to the survey, would pick a union, while the second choice is "a professional association," with 30 percent. The group activity of greatest interest to these professionals is "meeting with management to discuss policies."

The study largely reinforces much of what CWA is already doing to be attractive to these workers. However, it also suggests that new approaches need to be looked at.

In order to attract those workers who support "professional" kinds of employee organizations, we need to have the flexibility to meet those objectives with perhaps a form of organization that is, at the outset, less than full traditional unionism. This could be the first step toward a slow but steady conversion to full union representation.

The study also shows there is a large number of "fence-sitters," who express concern about conflict, shy away from a confrontational style, and worry that unionization might cause tension at work. Thus, to attract the commitment of those who are cautious supporters of collective action, unions of professional and technical workers will need to allay those concerns by vigorously demonstrating a willingness to solve problems by bargaining as equals with management.

I believe these findings by Professor Hurd are right on target. CWA organizers will make use of it as we move aggressively to recruit this new workforce and position ourselves to be responsive to a growing and diverse America.

While other surveys show that these workers may have a distaste for confrontation, when organized and mobilized in a union, they can be as militant as anybody else. Professional and technical workers seek an organization that will enable them to survive through an increasingly fluid economic time and achieve some sense of stability in their occupations.

In response to these needs, unions that want to represent professional workers may need to evolve into hybrid organizations combining the best of a trade union and a professional association. The union will provide the power on the job through the collective bargaining process, helping to improve wages and working conditions and creating a vehicle for workers to air grievances and gain a real voice in controlling their work.

But the unions may also need to look like a professional association, providing networking activities, career enhancement and development, certification or educational accreditation and other professional opportunities.

CWA is uniquely poised to be the union for professional, technical and administrative support workers in the 21st century. CWA is a union that has historically stood on the frontiers of technological change. Our members work in telecommunications, cable television, publishing, printing, broadcasting, health care, law enforcement, higher education and government. They are building and servicing the global information infrastructure. They are software and network designers. They are creators and disseminators of content. They are operators, customer service and sales representatives.

Growth and inclusion have long been the basis for our union's success. Today, our members work in all of the key jobs in the private and public sector that will make up tomorrow's knowledge-based economy. We no longer see the future based on a single industry, but rather on broad occupations where the skills, training and concerns of the workers cut across the boundaries of a single employer.

CWA has pioneered worker education programs, flexible work schedules, family benefits, worker empowerment, diversity and the strong community-based presence that appeals so greatly to professional and technical workers. As a result, CWA has been successful in attracting thousands of these workers to our ranks.

We have been strengthened by our mergers with the International Typographical Union, the National Association of Broadcast Employees and Technicians and the 35,000 member Newspaper Guild.

In 1996-97 alone, CWA organized 18,000 law enforcement officers in Florida; 5,000 other law enforcement workers; 3,500 research scientists at the University of California; 2,200 reporters, clerical workers and photographers at Dow Jones; and 9,500 customer service representatives at US Airways.

One of the biggest areas of employment for professional and technical workers is in government. CWA has been successful in organizing public workers going back to the 1960s in New York, when the Municipal Management Society affiliated with us. They represented managerial employees in all city departments and agencies. This was CWA's first big step into public sector organizing.

In the late 1960s and through the 1970s we organized virtually all the welfare workers in the state of New Jersey. In 1981 we won the election for 36,000 New Jersey state workers that included workers ranging from entry level clerical jobs to civil engineers, teachers and nurses. We have since added city and county workers in that state. Other public sector organizing has intensified and right now, more than 100,000 CWA members work in the public sector, most of them in professional, technical and administrative positions.

While a higher percentage of public workers are organized compared to the private sector, there are still millions of public sector workers without union representation. Many belong to smaller organizations or unions that could benefit greatly from a merger or partnership with a union such as CWA. With only 37 percent of public workers organized, we still have a lot of work to do.

The increase in the numbers of organized professional and technical workers has not been lost on those who would prefer that labor representation did not grow within these occupations. There has been an ongoing battle in Congress and before the National Labor Relations Board to redefine the legal rights of professional workers under the nation's labor laws.

Some four hundred doctors in southern New Jersey were rebuffed by the National Labor Relations Board in 1998 when they sought to unionize in order to bargain with the managed care corporations in their area. The managed care organizations have the power to set prices, but the doctors have little ability as individual practitioners to bargain effectively with them over fees and patient care. I encourage physicians and other health care professionals to persist until the law finally allows them the same rights as other workers.

There are ongoing attempts in Congress to make it easier for employers to classify workers as "independent contractors," which exempts them from coverage under the National Labor Relations Act and other protective laws. Another effort would have removed more workers from coverage of the nation's overtime law, even allowing employers to dock the pay of workers now exempt from the law (such as professionals) who, though they may work more than forty hours in a week, take some personal time off during the week.

One controversial piece of legislation would define a temporary employment firm as the "employer" for tax and benefits purposes, hence eliminating the real employer's legal obligations to these workers under the law. This approach would encourage the already explosive growth of temporary and "leased" employment arrangements which makes union organizing very difficult and generally results in lower wages and benefits to workers.

Despite the number of Americans anxious for jobs in the growing high-technology field, employers are using the excuse of a job shortage to expand the number of foreign workers who are allowed into the country to fill these positions.

According to the Labor Department, employment in the computer and data processing service industries would double between 1996 and 2006. This is impressive job growth. But one may wonder: Where will all the bodies come from to fill these jobs? Employment in this sector has already doubled during the past ten years without a recourse to major increases in foreign information technology workers being permitted to enter the U.S.

Labor is not against immigrant workers, but we are opposed to relaxing immigration standards so that foreign workers can take U.S. high-tech jobs when there are plenty of American workers who, if properly trained, can do this work.

Information employers are indeed experiencing growing pains that will undoubtedly continue in the future. But we believe that a legislative solution that encourages a dependency on lower-paid workers from overseas is not in the best interests of our nation or professional workers.

Organized labor stands ready to work with government and the industry to undertake an extensive training effort that taps the considerable wealth of human resources already available in this country. In the meantime, we will continue to work against legislative proposals that will allow greater numbers of foreign "guest" workers into the country at a time when American workers should have the training and opportunity for these jobs.

It is true that today's stars of the information industry are generally not represented by labor unions. Software writers, engineers, marketing and other professionals in the various information services might be making good money now, but what is in store for them in the future?

They may be young hot shots today, with their prime years ahead of them and convinced that they can survive on their own. But once they see how power dominates talent, they may think differently. Many already have, as evidenced by the successful legal challenges to Microsoft's excessive use of temporary and contracted-out workers.

In many of the creative fields that the information age is making so profitable, such as graphic arts and writing, freelance professionals have little or no protection over their intellectual property rights. When a freelance writer submits an article for publication, the publisher can demand to have sole ownership rights.

The writer or artist has little bargaining power. A publication can distribute their work product on its web site, resell it or reprint it as often as they like without paying another cent to the creator. Graphic artists who are not represented by a union can see their work altered or used in ways they had not intended, in addition to being repeatedly sold for a single fixed fee.

CWA is working on issues such as protection of intellectual property rights, particularly as they pertain to the Internet, in order to protect the products of creative professionals. We believe that this is one of the most important labor issues facing the professional worker.

Also shaping the new American workforce is the dramatic shift in demographics. A majority of those entering the labor force are women, including a large number of African-Americans, Asian-Americans and Hispanics. The white male is already in the minority. By the year 2050, more than half of the American workforce will be comprised of ethnic minorities.

Women workers are absolutely vital to the future of the labor movement. Since 1970, women have gone from nearly 22 percent of the total number of union workers to more than 39 percent. Yet, some 87 percent of women workers do not belong to unions and a majority work in professional, technical and service jobs.

The AFL-CIO is doing an excellent job of reaching out to women workers, but we must do better. Labor must demonstrate through word and deed that women are welcomed and encouraged to be part of our ranks.

CWA union women are among the best in the labor movement. Women were the backbone of this union when it was first formed in 1938 and they remain the backbone of CWA today. Women comprise about 52 percent of our membership.

Issues involving work and family receive attention at the highest levels of this union. Job stress, family matters, flexible work schedules, maternity leave, the special health needs of women, sexual harassment, equal pay, career development and education are issues at the top of CWA's agenda.

CWA has done a solid job of opening doors of opportunity for our minority and women members. In fact, a union job is one of the best affirmative action programs in the nation.

Unionization has a dramatic impact on women's wages. Wages for women are historically lower than those of men. In 1997, Heidi Hartman, director of the Institute for Women's Policy Research and a MacArthur genius grant recipient, analyzed the impact of collective bargaining on women's wages. She found that unionized women's wages were $2.50 more per hour than nonunion wages, and after controlling for all variables, such as education, industry and firm size, Hartman concluded that collective bargaining adds 12 percent to women's pay.

Our union is determined to make sure that our members, regardless of their gender or race, receive equal opportunity and justice on the job and in the communities where they live.

We are just as determined that CWA reflect the many faces of our union in the makeup of our professional staff. I am committed that we will achieve that goal before I hang up my spikes as president.

We are making significant progress. In 1994, 28 percent of the CWA staff were women. Today, 35 percent are women. Women have filled twenty-three of the twenty-seven additional staff jobs that have been created over the past three years. There are now eighty-two women professional representatives, compared to fifty-nine in 1994.

Our minority hiring has improved as well. From 1994 to today, we have gone from 15 percent minority staff representative to 17 percent, from thirty-one jobs to forty-one. Our numbers are even better in the national headquarters. Our staff in Washington is 45 percent female and 18 percent minority. So we are closing in on our objective.

In CWA District 9, which covers California, Nevada and Hawaii, we achieved a historic breakthrough in 1998. For the first time, a CWA district has more female staff than males. In District 9, fifty-nine percent of the district staff are women. CWA Vice President Tony Bixler deserves our thanks for his commitment to diversity and equal opportunity within our own ranks.

Unions are political organizations. Historically, appointments to staff involved political considerations, particularly when the vice president continually had electoral opposition.

Glenn Watts, when he was secretary-treasurer, began to push President Beirne into an affirmative action program. It was quite testy for a while, but in 1972, Joe got us all to buy into a program that would put more women and minorities on staff. He told each of us that our next vacancy had to be filled by a woman.

As luck would have it, I had the first vacancy, and recommended the vice president of Local 1177, Larry Mancino. Joe came down hard on me, and all I could think of saying was, "Joe, are you prejudiced against Italians?"

Joe relented. Larry went on staff, rose in the ranks to New York director, assistant to the vice president, from 1991-96 as my assistant in Washington and then was elected vice president of District 1. He is one of the outstanding trade unionists in CWA.

Today, I am fortunate that I have the commitment of all our officers to shape the future of our union. We all look forward to the day when the CWA Women's Committee and the Equity Committee are history, something that was very much needed in our past, but which achieved its mission insofar as equality of opportunity within CWA is concerned. My personal goal is to still be around when that day arrives. Organizing the new workforce in large numbers will be more difficult for us unless we resemble those we want to join our ranks. We are almost there.

We have proven repeatedly that CWA is a union that appeals

to the new American workforce. These workers share common hopes and dreams whether they are telephone operators, computer technicians, customer service representatives, college employees, government workers, police officers, broadcast technicians, sales people or journalists.

The labor movement must continue organizing service and retail employees. These jobs will never be automated out of existence and those doing the work need union representation more than ever. The twin strategies of targeting both professional and service workers complement each other. The more that unions represent workers at the high end of the wage scale, the more power it will have to protect those at the bottom.

In a time of great change and uncertainty, the new workforce is probably more open to our message of union organization than at any other time in our history. Advanced technologies are rapidly changing their working lives. As we shall see in the next chapter, the convergence of information technologies is creating a digital revolution that will have an impact on human life that the world has not seen since the invention of the printing press some five hundred years ago.

Chapter Fourteen

Convergence

When I worked as a telegraph operator, we thought of ourselves as indispensable. The human skills required to transcribe Morse Code to English text, reading through sunspots and static, and having the sender resubmit portions of the message that were illegible made us feel that our jobs would always be secure.

We were able to do what few others could. We felt certain that we would never be automated out of existence. We were wrong.

Technological innovations developed during World War II helped stabilize the radio signal and allowed it to be put directly on teletype. I remember when the first teletype machine entered our workplace, and we all said, "that thing is not going to replace us." But it did.

Fortunately, we were CWA members by then. We negotiated the first automation clause in any industrial service sector collective bargaining agreement. The contract protected our title, wage rate and seniority. We kept our jobs while the company had flexibility to assign us to other work. Without the union, we surely would have been out on the street.

Today, many workers face similar challenges. We stand at the threshold of a new information age, participants in the most powerful technological and information revolution that human-kind has ever seen. Every aspect of our lives and work will be affected by the development of information technologies.

One hundred years ago, at the end of the 19th century, human thought could be expressed in dots and dashes, the language of the telegraph key. Today, at the turn of the 20th century, nearly all of human creativity can be translated into bytes of information, the 0s and 1s of the digital revolution.

Nations fought wars over oil in this century. Tomorrow, information will be the prized commodity of exchange. Who produces information, owns it, distributes it and controls it will become the oil barons of the 21st century. And whoever controls the means and infrastructure of data transmission also controls our work. The lives and jobs of workers and their families will change in ways that we cannot yet imagine.

We need to understand the changes that are taking place and how they will affect the industries in which we work. Since CWA members are employed by some of the leading information and communications companies, we are among the first workers affected by these trends. One of the most significant developments is what we call convergence: the industries of telecommunications, information, media and entertainment—TIME for short—that once existed as separate and distinct industries are now converging into one giant information industry.

Information technology has already merged what used to be distinct industries into one large information sector. Telephones, computers, broadcasting, publishers, cable and other communications industries were up until recently considered separate industries. Now digital technology brings them together into a single industry. Telecommunications facilities can carry television signals.

Cable and direct broadcast satellite television are available in 95 percent of America's homes. Publishers own radio and television networks and cable systems. The final shape of this new industry has yet to develop and there are many visions of how it will ultimately take shape. But dramatic changes are already occuring.

Broadcasters, cable television, computer makers, publishers and other information producers each have their own visions of how Americans will receive, use and send information in the future. The FCC has mandated that by 2006 over-the-air broadcasters must convert to digital technology. Analog broadcasting as we know it today is history. Whether the complete switch over actually takes place by that time is uncertain, so don't throw away your television set. But digital technology is reshaping how we send and receive information.

Microsoft, for example, sees your television evolving into a fully interactive computer able to surf the Internet with voice commands and download whatever program or information you need on the fly. Broadcasters envision wall-sized, flat television screens showing the latest movies in true to life colors and CD-quality sound. Cable operators are out to control the format of the box that will sit on top of your new communications set (it may no longer be called a television) through which signals will pass.

We just don't know at this time exactly what the future holds for us as workers and communications users. In anticipation of these great changes, we already see the incredible rise in the value of data transmission as the commodity of exchange in the converged technologies.

The demand for Internet and data services is exploding. By the year 2010, more than 95 percent of all businesses will be using data networks, compared to approximately 50 percent today.

Sol Trujillo, president and CEO of US WEST Communications, says that by the next decade, data will represent fully 80 percent of the traffic on his company's network and voice will only be 20 percent.

GTE predicts that by 2006, voice traffic will decrease to 28 percent of total earnings, from 52 percent today, and data will grow 10 to 13 percent annually to 53 percent of total revenues. All of the major telecommunications companies have made similar projections.

As the information industries converge, so must their unions. Unions have been talking about and preparing for convergence throughout the 1990s. In one of our first merger meetings, former Newspaper Guild President Chuck Dale and I recognized how our members were going to be affected by this new technology. The members of The Newspaper Guild are among those who produce information. CWA members in telecommunications transmit and maintain the systems that distribute information. Our members also help users navigate the new information systems.

We represent many common occupations. CWA members do the same advertising, sales, circulation, customer service, clerical and administrative jobs performed by Guild members. And we are all impacted in the same way by the same technology that is blurring the job distinctions between inside plant and outside plant, editorial and production, professional and technical.

The far-sighted leaders of The Newspaper Guild had long seen the transformation taking place in communications and the profound impact of new information technologies on our jobs. The global change in how news is gathered and distributed, new production technologies, the growing concentration of media ownership, advertising and circulation challenges were all things that

TNG members experienced on the job long before the rest of the public was even aware of the information revolution.

Our merger was a natural partnership. It was only a matter of time before the major union that represents those who distribute information joined together with the most important union that represents those who produce information.

No union merger probably has ever gone as smoothly as the one between CWA and TNG. I asked Chuck to write down a list of things that the Guild would need to merge with us. When he did, I took one look at it and said, "We accept." While it wasn't all quite that easy, we later agreed that we never expected it could go so smoothly. As smooth as a merger goes, it is always traumatic for the smaller union. We work hard to make them feel right at home.

Like the Guild, CWA has been forced to take a different view of ourselves and how we respond to the new challenges. We saw that the rise of global competition, deregulation, advanced technologies and changes in media ownership rules were irresistibly driving the convergence of the telecommunications, information, media and entertainment industries.

We came to the conclusion that no one union, acting alone and independently, could adequately protect or defend their members under these tremendous pressures. We made a conscious decision to become a new and different union. Our partnership with the Guild is a significant component of that transformation.

Unfortunately, many publishers dream of operating in a union-free environment. This requires us and other unions at a newspaper to develop a sound strategy before bargaining begins. In July 1995, our members in the composing room, along with those of other unions, went on strike for a fair contract at the *Detroit*

News and *Free Press*. These papers are owned by the Knight-Ridder chain and Gannett Corporation and do business under a joint operating agreement.

Early in the strike, the companies began to hire permanent replacements for the strikers, demonstrating they would rather try to bust the unions than negotiate in good faith. Following a back-to-work order by the unions that was ignored by the companies, the strike became a lockout as a third year was concluding.

How do workers survive for so long? Our members in the composing room received $200 a week from the national union and $50 a week from our printing sector. The Guild was not merged with CWA when the strike began and paid $150 a week strike benefits to its members.

We began an Adopt-a-Family program where we would give families in need, including Guild members, an additional $500 a month. Several hundred CWA locals, staff and officers participated. More than $2 million was contributed and every family in need was adopted.

This demonstration of solidarity was a two-way street. During our 1996 convention in Detroit we had an enormous rally at the struck papers. Being in Detroit gave our local delegates an opportunity to meet and visit with their adopted families. Several delegates, with great displays of emotion, spoke on the convention floor about their visit to the homes they were helping to keep intact.

I am extraordinarily proud of the many locals that contributed generously. They demonstrated that CWA is, indeed, a family union. When one of us hurts, all of us join together to ease the pain.

This experience has also made it clear that new strategies are required to deal effectively with the major newspaper chains.

We do not have to prove how militant we are by calling a strike. We need to deal with the reality that striker replacement is a weapon that sits silently on the employer's side of the table. There are many powerful alternatives to strikes and we need to work them into our strategic planning. We proved this, for example, in our five-month struggle with Bell Atlantic in 1995.

The members of the National Association of Broadcast Employees and Technicians (NABET) face similar challenges. They felt the impact of concentrated media ownership when General Electric bought NBC. Almost immediately, what had been a stable relationship between labor and management for decades was thrown into turmoil.

Riding the wave of antiunionism in the 1980s, GE forced NABET to accept a series of concessionary demands that culminated in the 1989 strike where our two unions first began working together. They realized that a smaller union such as theirs couldn't possibly match the resources of a multi-billion dollar corporation such as GE.

Since its merger with CWA, NABET has been working hard to organize the unorganized radio and television stations. As the industry concentrates into the hands of fewer owners, such as Capital Cities buying ABC and then Disney acquiring it all, it has become clear that new strategies need to be developed that can exert pressure on the entire organization.

As evidenced by the existence of these media giants, convergence is clearly a global phenomenon. Every nation is now or will be affected by this technological revolution. In early 1998, the European affiliates of Communications International, the international secretariat for telecommunications and postal unions, met to discuss convergence.

They concluded that organizing must be at the top of their agenda. They are committing themselves to organizing both in the existing industries and in the new industries that are forming as a result of convergence.

This is a very new approach for the Europeans. The media, information, telecommunications and entertainment unions in Europe are also looking hard at mergers among themselves. Indeed, convergence is one of the driving forces that is leading to a proposed merger of several of the international secretariats that represent our members.

Mergers with other unions that share our vision for the future certainly make sense for CWA. We now have ongoing merger talks with several unions and other cooperative efforts and will pursue this strategy in the future. Not every discussion may result in a merger, but we can certainly find ways to work together and assist each other in organizing, collective bargaining and political action.

In 1997, for example, CWA and the American Postal Workers Union opened discussions. We have engaged in joint organizing activities and, in 1998, we agreed to joint political action. In July 1998, I was honored to speak at the APWU's convention and was received very warmly by the delegates. We have extended an invitation to Moe Biller, president of the APWU, to address our convention in 1999.

Ironically, this is not the first time that we have talked about getting together. Back in 1971, CWA, the APWU and the Letter Carriers actually agreed to a structure for a new union. But before it was enacted, the Letter Carriers backed out of the proposal. Here we are again, almost three decades later.

What brought renewed interest in closer ties between CWA

and the APWU? Convergence most certainly is bringing us back to the table. Clearly, our two industries—telecommunications and postal service—are being drastically impacted by the convergence of information technologies. We face the same threats of growing competition and privatization. E-mail, the Internet and other technological advances on the horizon that we can't even imagine today will affect our jobs in the near future.

How much longer, for example, will the postal unions on their own be able to withstand the repeal of the private express statutes that now guarantee the United States Postal Service a monopoly over first class mail?

We have the same problems with contracted-out and temporary workers taking our members' jobs. We share the same concerns over health care, retirement, Social Security and other workers' issues.

The APWU and CWA have a long history of cooperation. Our national leaders have worked closely together in the AFL-CIO and in the international labor movement. We share a common culture of strong local unions, grass-roots democratic traditions, and a heritage of progressive labor policies. If a merger takes place between CWA and the APWU, it will be a bottom-up process, with local leaders and members in the forefront.

I don't know if our current efforts will result in a merger. But we are at least trying to make something happen. It is simply not enough for us to circle the wagons and just take care of our existing members. We must organize the entire industry if we are to survive and grow and continue to protect our members. Unions need to combine their strengths to better represent their members in the face of mergers in which our employers are now involved.

Our employers understand the power of mergers as convergence turns the laws of classical economics upside down. If a manufacturing company grows by 100 percent, it does not necessarily follow that it will increase profits by 100 percent. That is called the law of diminishing returns and it applies to most business endeavors.

Convergence, however, is opening up a new world of knowledge-based products and services. In this new world where our members work, innovation and size can lead to ever-increasing returns, contrary to the law of diminishing returns.

Mega-mergers, joint ventures and the drive to get bigger and bigger are key business strategies to higher profits in the converged industries. In this world, the law of diminishing returns from increased growth does not apply. Mergers, such as Disney-ABC, Bell Atlantic-NYNEX, SBC-PacTel-Ameritech, and Rupert Murdoch with just about anybody, all make good economic and business sense. Whether they all serve the public interest, however, is another matter.

That is why CWA is so concerned about the proposed MCI-WorldCom merger. This is a move that threatens consumers and workers. Although both companies are known for their long-distance businesses, the MCI-WorldCom merger is not about long distance voice communications. It is about who will control the Internet.

We oppose the MCI-WorldCom merger because it is not in the public interest, the interest of the industry, our nation or our members. When I first read about the deal, I was shocked. How could one company buy MCI just by putting up paper? This is one of the largest mergers in history—a deal valued at some $37 billion. Yet, WorldCom is using its stock to pay for 80 percent of MCI. A little revolution somewhere in the world, another Asian collapse, or an interest rate hike by the Federal Reserve, could cause shockwaves on Wall Street. Along with it goes WorldCom's stock value. WorldCom is a house of cards waiting to fall.

The news media tells the story that just before they closed the deal, MCI CEO Bert Roberts asked WorldCom CEO Bernie Ebbers for just a little more to sweeten the pot. "Sure," Ebbers reportedly said. "Instead of $50 a share, let's make it $51." Well, why not? It's just paper. This is a world that few working families will ever understand.

What is even more distasteful is the obscene millions that the top MCI executives will receive if the deal goes through, all at the expense of shareholders and consumers. For supporting the WorldCom merger, the top corporate officers of MCI stand to personally get millions of dollars.

As this book goes to press, I don't know the outcome of our campaign against the MCI-WorldCom merger. When we first began the effort in 1997, experts thought we were tilting at windmills. The merger may go through, or it may not. But we have already forced significant changes in WorldCom's and MCI's internal strategies. MCI has provisionally sold its Internet business to Cable & Wireless to make the merger more palatable to regulators in the U.S. and Europe. But this should not be enough for approval.

Regardless of the outcome, we have made the voices of our members heard in this debate and that is a very important objective for us. Needless to say, the business press roundly criticized our involvement. *The Wall Street Journal's* editorial page attacked us. MCI President Timothy F. Price accused GTE of "bankrolling" our campaign. His charge, of course, is a lie. But we must have done something right to spark such an intense reaction.

Our efforts in the MCI-WorldCom campaign demonstrate the role that unions must play in the future to have an effective voice in the policy issues affecting our members and our nation.

CWA was deeply involved in the development of the 1996 Telecommunications Act and its passage. We believe that the Act will expand access to advanced telecommunications services through lower prices, stimulate investment in new facilities and equipment, improve service quality, speed deployment of new telecommunications technologies, and promote the growth of good jobs in the industry.

But none of the promised benefits of the information age are automatic. Just as the industrial revolution created vast new wealth alongside devastating poverty, the information revolution holds the same dangers. Millions of our citizens could be left behind, thus further aggravating the already wide disparities in income and opportunity. On a global scale, the potential is even more frightening.

I had a discussion with Mexico's Minister of Labor about the future power of computer-driven technology to improve life for our children. I used the analogy of two children, one an American growing up in Brownsville, Texas, and the other a Mexican in Mexico City. I spoke eloquently (so I thought) on the need for both of them to have equal access to the new information technologies so that they would be able to compete in the new global economy. He agreed with me completely. But then he stretched out his hands, palms up, and said, "I have so much on my plate right now, I'm just trying to make sure that the child in Mexico City has clean drinking water."

Nearly half the world's population has never made a telephone call. When I first told this to President Clinton, he couldn't believe it. He turned to Vice President Gore and asked, "Al, is that right?"

"Yes, it is, Mr. President." Gore responded.

Recently, the CWA Executive Board adopted a statement that outlines our policy priorities as we move forward in the development of the information industry. Our priorities include: Promote universal service; encourage growth and investment in the network; provide a level playing field for all competitors; and foster the creation of high-wage, high-skill jobs.

The goals of universal service must be balanced with the desire to create competition. Universal service and access are often used interchangeably, but they are distinctly different. Any of us can walk into a Rolls Royce dealership and sit behind the wheel. We have access to the car. But how many of us can afford to drive one home?

Universal service means that the fruits of the technology be readily accessible to all citizens at an affordable cost. Neighborhood libraries should be a place where all information can be accessed without a fee.

Our democratic society cannot afford to have cost restrict our citizens' ability to participate in the information revolution. Our government has historically promoted the free flow of communications, ideas and speech.

In the old days, we used to say that you can't argue with a person who buys printing ink by the barrel. In today's information rich society, anybody can be a publisher. With the touch of a keyboard, you can "flame" anybody you disagree with.

The new information technologies should encourage greater interaction and communication. But there is an enormous gap that must be closed if ordinary citizens are to participate in the information revolution.

We must insure that our society does not become divided between those who have access to information and those who do not. Otherwise, we risk exacerbating the spread of wealth and opportunity that already exists in our nation and the world.

CWA is committed to doing our part to encourage universal service. Our members have volunteered to help wire our nation's schools in all of our poorer school districts where our employers contribute the equipment. We will work with school boards and our employers to safely wire and install the equipment for our schools to have access to information services.

CWA members have actively participated in the annual NetDay initiative begun by Vice President Gore in 1996 to wire the nation's schools. By 1998, CWA members had joined other highly skilled volunteers from AFL-CIO unions to wire seven hundred of the nation's poorest schools to the Internet. We are very proud of that achievement.

Telecommunications policy must also reinforce the philosophy that success in the new telecommunications marketplace must be built on superior technology, service and content, not on depressed labor costs that can lead to the deterioration of service quality.

As CWA members learned, competition in long distance did not live up to the promise of good-paying, middle-class jobs. From 1984 to 1996, we estimate that 115,000 jobs have been added at the new companies, compared to the 124,000 jobs that AT&T has shed. And the new jobs pay significantly less than the jobs that were eliminated, so the average earnings in the industry have declined.

For example, a service representative at Sprint Long Distance earns two-thirds of what a service rep at AT&T earns. A

Sprint technician earns 85 percent of the earnings of his or her counterpart at AT&T.

The wage gap doesn't tell the whole story. Workers in the new jobs receive fewer benefits, pay more for their health care and can look forward to a smaller retirement check, if they get one. We have had to fight every step of the way to protect our union-won wages and standard of living.

This low-wage, low-skill threat is the underside of competition that the public rarely hears about. The new jobs of the future could be high-wage/high-skill. But without an explicit public policy, most of the new jobs will be low wage, part-time, temporary and contracted out.

I don't think this is a future that any of us wants.

With no clear direction toward the growth of good jobs, the information revolution could take a wrong turn. CWA is committed to fighting for high-skilled jobs that provide for secure employment opportunities, decent benefits and a hope for a secure retirement.

In 1995, I was appointed to the President's Advisory Commission on the National Information Infrastructure. During the initial meetings of the Commission, I became greatly concerned that the interests of Wall Street were driving the discussions.

The first outside expert invited to speak to the Commission was an executive vice president from Merrill Lynch who follows the telecommunications industry. He told us Wall Street's vision of the information age. They have us programmed. He predicted that Americans will watch six videos a month at $4 each, spend some $300 to $400 billion on home shopping networks, and, in the time left to us, we will play video games. All at an enormous profit to information providers.

I thought that it was highly symbolic that of all the academics, industry experts and public policy makers we could have invited to be the first outside speaker to the group, an executive from Wall Street was chosen. This choice reflects the power of Wall Street in shaping the future vision of the information age. Wall Street sees the information superhighway as the fast track to the bottom line. And Wall Street's vision is driving the companies to concentrate on the development of the high-profit, high-margin services and products.

Along with several other members of the panel, I expressed apprehensions about the direction of this discussion. The final report, which also influenced the Telecommunications Act, addressed our concerns that we also focus on public interest issues such as insuring access for the disabled, education and other matters.

The President's Commission was just one step along the way in our national debate over the future of the new information technologies. Because CWA stands on the information frontier, our members can see the directions that advanced technologies are taking us much more clearly than others. Our union has a responsibility to sound the warning bell when we believe developments are taking us in the wrong direction or when public policy is needed to correct abuses.

CWA is taking an active role in the development of public policy, shaping the debate and identifying solutions. Along with the AFL-CIO, we are fighting for an information age future that leads to high-wage, high-skilled jobs in high-performance workplaces.

I serve as chair of the AFL-CIO's Technology Committee, which was formed to develop a unified labor policy on issues that concern working families in the information age. At the August

1997 meeting of the AFL-CIO Executive Council, the committee issued its first report. We concluded that it was vital for the labor movement to develop a broad policy on the national information infrastructure and pave the way for trade unionists to become active in every level of its development.

I am reminded of two visions of the information age.

Early in the Clinton-Gore Administration, Vice President Gore expressed his vision of the information superhighway. He said, "I want a child in Carthage, Tennessee to be able to access the Library of Congress in Washington, D.C."

Several months later, the president of one of the RBOCs gave his company's vision of the superhighway. He said, "Imagine a mini-theater in every family's living room in the U.S."

Which of these two visions of the information age will prevail? That is the question that every American must ask. As workers and citizens, we dare not leave it to others to answer that question for us.

Like the industrial revolution before it, the information revolution has the power to bring people all over the world closer together. The global village is a reality. As a result, our international labor alliances are more important than ever.

CHAPTER FIFTEEN

International Unionism

My first exposure to the dangers faced by foreign trade union leaders occurred in 1970. I had been vice president of District One for a year and had been invited to attend a meeting of the Inter-American Committee of the Postal Telegraph and Telephone International. PTTI, which has since changed its name to Communications International, is the international labor secretariat that represents some 4 million telecommunications and postal workers all over the world.

The meeting was held in Cuernavaca, Mexico. At one point during the meeting, an anti-Castro resolution was being debated. A labor leader from Venezuela, who represented the telephone workers there, said that before he was granted a visa to go to Mexico, he was told by his government that he may not be allowed to return if he participated in any anti-Castro activities. He then spoke on behalf of the resolution, defying his government's political threats.

I left that meeting with great admiration for his courage. Since then, I have met many labor leaders from around the world who have faced death threats, police harassment, and been abducted in the middle of the night and taken away from their families for months. During their captivity, they experienced all kinds of unspeakable torture and deprivation. But they never gave up.

For many, the support they received from the AFL-CIO, PTTI, and unions such as CWA was the only thing that stood between them and a lifetime of imprisonment or even death.

The international labor movement has played a significant role in the tumultuous events of the closing two decades of this century, as democracy spread around the world. We have seen the fall of the Soviet Union and the emergence of democracy in Eastern Europe, the opportunity for peace in the Middle East and Northern Ireland, and the defeat of apartheid and the triumph of democracy in South Africa. International labor was prominent in all these events.

The fall of Soviet communism began in 1980 in Poland when an electrician by the name of Lech Walesa climbed over a fence in the Gdansk shipyard to lead a strike. For nearly ten years, the Polish union Solidarity worked underground to bring democracy to Poland. They were supported in their struggle during all of that time by the American trade union movement.

Solidarity's victory was the first break in the Iron Curtain— eventually the entire Soviet system collapsed. I remember being in the White House with great pride as a labor leader when President George Bush awarded the Medal of Freedom to AFL-CIO President Lane Kirkland and Lech Walesa. I find it ironic that Ronald Reagan supported Solidarity in Poland, but fired the air traffic controllers here—he supported a free and strong labor movement, so long as it was not in America.

Now that so many countries throughout the world are moving toward free market economies, the role of the international labor movement is more important than ever. A free trade union movement is essential to democracy. Whenever a dictatorship takes over, the first thing they do is lock up the trade unionists.

We saw that in Germany in May 1933, where the first inmates of Dachau were union leaders.

Chinese authorities tolerated the democracy movement as long as it was confined to students. When workers joined the protests, the tanks rolled into Tiananmen Square. From the tragedy of Tiananmen rose the formation of the Chinese Workers Autonomous Federation, which is now working underground to bring democracy to China.

One of my colleagues, an Argentinean unionist, Hernol Flores, drove a mail truck during the days of the military dictatorship. He came to the Executive Committee of the PTTI (now Communications International) in Santiago, Chile to brief us on the situation in Argentina. In his role as labor leader, Flores fought the military dictatorship, often risking his own life and those of his family. He showed us photographs of his automobile riddled with bullets. He had newspaper reports as well. His would-be assassins—members of the military—had mistaken his son for him, and shot up the car his son was driving. The car crashed as a result of the tires being blown out, but luckily he was not killed. This had happened just a few days earlier.

We offered to take him and his family to safety in Canada or the U.S. He told us, "No, I've got to go back to the fight."

When a democratic government was finally established in Argentina, Flores was offered the position of labor attache in Washington, DC. Despite the fact that this was a very prestigious job, Flores turned it down, saying that he had to stay in Argentina and rebuild his union.

Sometimes we have to put our own struggles, as important as they are, into context. Following a public hearing on the La Conexion affair conducted by the Labor Department, one of the fired workers who had just testified was asked by a news reporter: "If you knew you could lose your job, why did you keep supporting the union?"

"What does risking a job matter?" the woman, a native of Peru, replied. "In my country, workers have risked their lives to have a union."

I have often asked myself, would I be a trade union leader if it meant risking my life or my family's? If I did take those risks, would my wife stay with me? I honestly do not know the answer to those questions. I do know, however, that there are a lot of heroes all over the world who have risked their lives fighting for the trade union movement and a democratic society.

Politicians, and even governments, come and go while the international trade union movement remains. The movement is built on solidarity and unity of purpose, the commitment to give everybody a chance at dignity and quality of life. A trade union movement speaking in a single voice in every country transcends governments and national borders. It is the most powerful common interest joining working families around the world.

As we in America continue to debate the role of labor unions in our country, or even whether they should continue to exist at all, other industrialized nations have long embraced trade unions as full and equal partners. Strong labor movements are the rule, not the exception, in the nations who are our toughest international competitors.

In some countries, the unions are reorganizing in order to remain powerful in the competitive times ahead. The German trade union leaders who survived incarceration by Hitler believed their labor movement was too weak and fragmented to resist Nazism.

I spoke with a German trade unionist who was a survivor of Hitler's purge. He said the general belief was that there were too many unions. They resolved that in post-war Germany it would be different.

And it was.

Post-war Germany had fewer unions and it wasn't long before the leadership believed they had to consolidate even further. It was not necessarily driven by their desire to have the labor movement be a bulwark for freedom, as much as the need to deal with more powerful transnational employers.

In February 1998, I was at the Free University in Berlin making a presentation at a symposium. That evening I was briefed by my German colleague and friend Kurt van Haaren, president of the Telephone and Postal Workers. He was attending a meeting where seven of the fourteen German trade unions were planning to merge.

The mergers of the seven unions will establish one union for all service workers. Within that broad union there will be various sectors, which will have a degree of autonomy in order to maintain identity and avoid jurisdictional disputes. The merged unions will have the ability to represent workers ranging from entertainers to technicians, professional and low skilled jobs—all broadly defined as service workers. The union will be a powerful voice for millions of workers.

Throughout the European Community, labor unions are accepted as part of the social, economic and political life. They are recognized as legitimate and necessary organizations, vital to their nation's health. Even in the European countries where union organization is not so highly concentrated, the antiunion animus prevalent in America simply does not exist. European workers do not have to worry about striker replacement or company campaigns against union organizing. Such practices violate the International Labor Organization conventions. Of course, the ILO conventions do not apply here. The United States sends representatives to the ILO, but the U.S. Congress doesn't ratify the conventions.

Not only should the United States live up to standards of labor practices followed by most other developed nations, but we also need to establish stricter international labor standards that every country must abide by. These standards should include: the freedom to join or organize a union; the absolute prohibition of child labor; an established minimum wage, whatever that wage might be; and finally that each country enforce its own labor laws. Mexico, for example, has better labor laws than the United States, but they are not enforced.

Over the past few years, a number of telecommunications unions throughout the world have been working together to develop international solidarity in the information industry. Known as the Working Group on Multinationals, under the auspices of the Communications International (CI), each union with membership in the multinational firms is included. The group has prepared the following code of conduct applicable to all firms operating internationally in the information industry.

1) Disclosure as soon as practical regarding the company's global activities, including forecasts of employment levels, possibilities for technological change, movement of work and investment decisions.

2) Annually and whenever else necessary, meeting with all organizations representing their employees globally. At such meet-

ings, general information shall be exchanged including organizational rights, equal employment opportunities, safety and health, and education and training.

3) There shall be no management interference in worker organizational efforts in any country where the firm or any subsidiary conducts business.

4) When presented with the appropriate level of organizational support relevant to a particular nation's recognition standards, the firm shall recognize the union and its representatives.

5) Production should not be shifted from nation to nation to seek low wages or avoid union recognition.

These standards should not only apply to every telecommunications company operating in the world today, but also be used as a model for every corporation in the global marketplace.

One of the most important ongoing international issues is the exploitation of labor throughout developing countries. We have a moral obligation to demand that our own government not do business with nations that exploit workers and children, or deny their basic rights to organize, strike or engage in collective bargaining.

Through the International Confederation of Free Trade Unions (ICFTU), international labor is pushing the World Trade Organization to accept core labor standards as part of the global economy. "The global economy needs global rule," the ICFTU said to the WTO. "The global market is a powerful mechanism for dynamic development, but it can also lead to the exclusion and marginalization of millions of ordinary citizens who do not have the advantage of wealth or status. It has to be balanced by countervailing powers. Achieving progress on international workers rights and trade is key to the future of the world trading system."

American labor certainly supports that declaration.

It is intolerable that America, the country with the longest history of democracy, is not more committed to the labor unions which help make that democracy work. Imagine if the Wagner Act had not been passed, the mass organizing drives had not taken place and the 1930s had been a time of even greater economic discontent. America could have just as easily turned to a dictatorship, and our country, as well as the world, would be a different place today.

Even a country as similar to the United States as Canada shows a completely different attitude toward trade unions, particularly in the past half century. Between 1955 and 1990, union membership in Canada grew from 31 to 36 percent. During the same period, union membership declined in the United States from 33 to 16 percent.

That is another important role of international unionism, creating a dialogue between workers and their representatives from different countries in order to educate them about how important it is for us all to stand together. We need to make workers understand that somebody else's fight today might be their fight tomorrow, so they are willing to stand up for each other across national borders.

Right now some insurance companies ship their records to Ireland, where data entry techs making a pittance transfer those records onto computer. If the Irish workers agitate for higher wages, then those jobs will go to Sri Lanka or somewhere else.

In the late 1980s, Trammel Crow, a large Texas real estate company, opened a facility in China. They taught Chinese college graduates the English alphabet without teaching them how to read English, and trained them to operate a computer keyboard. Reportedly, Pier One was the first customer and the Chinese facility owned by a Dallas real estate company paid Chinese workers a dollar a day to input Pier One's inventory onto computers.

Around the same time, American Airlines moved their billing center from Tulsa, Oklahoma to Barbados. In Tulsa, the company was paying American workers up to $13 an hour, plus benefits. In Barbados they only paid the workers $3 an hour with no benefits. Every day a plane landed in Barbados, carrying the company's records. The Barbadian workers punched in the data and it was transmitted back to America via satellite. At the same time, the company was fighting local efforts to organize the Barbadian workers.

A meeting was held in Canada, sponsored by the Canadian Auto Workers, for all the labor unions representing workers employed by Northern Telecom. The union president from Malaysia told us that he could not understand Northern Telecom's philosophy. His workers were making only $4 a day producing equipment, yet the company was so greedy it was bringing workers in from Bangladesh to work for less.

Many other labor leaders from all over the world had similar stories to tell. At the end of the second day of the conference, a woman from Ireland spoke up. She said that before the conference, she and her colleagues were grateful to Northern Telecom because it had brought desperately needed jobs to their region. But after what she had heard for two days, she realized why Northern Telecom had come to Ireland, and how Irish workers were being exploited.

As international corporations become less and less confined

by geographical boundaries and capital becomes even more mobile, the labor movement has to build stronger ties among unions internationally. Left unchecked, capital will find the cheapest labor market. The only way to protect jobs and wages, not only in America but throughout the world is with a strong international labor movement.

CWA is working very closely with our international allies. In 1997 we had three organizers in the United Kingdom for six months, working with British colleagues in the Communications Workers Union (CWU), organizing American companies in the cable and telephone industries over there. We have a close alliance with the CWU in the United Kingdom. To assist our Mexican colleagues, we held an organizing training school for them in El Paso, Texas. That training paid off with organizing victories.

In the converging telecommunications industries, the global market will shake down to a few major players with their own spheres of influence. The likely winners will be AT&T, British Telecom, Deutsche Telecom, the Japanese telecommunications giant NTT, and perhaps SBC and Bell Atlantic. Everybody else will either be swallowed up, or find their own little niche either within the industry or outside of it.

CWA already has a long history of working closely with our brothers and sisters in the telecommunications unions employed by these foreign companies. And those relationships have resulted in a stronger union presence at the bargaining table.

During our seventeen-week strike against NYNEX in 1989, we used an international strategy for the first time as part of our bargaining strategy. The results were remarkable. The PTTI adopted a resolution supporting our members and condemning NYNEX at its World Congress which was held that year in

Brighton, England. This was the first time we had asked PTTI to take an action in support of one of our disputes.

We targeted mailings throughout the world where NYNEX had foreign holdings, asking for the support of trade unionists and political leaders. We had many responses.

CWA Executive Vice President M.E. Nichols went to the U.K. to rally support, and as a result Labor members of Parliament introduced resolutions condemning NYNEX in the House of Commons. We went to the European Community about NYNEX and two commissioners came to the U.S. to express their concern to the company's management.

Akira Yamagishi, president of the telephone workers in Japan and leader of our international trade secretariat, CI, kept calling me regularly, to see how the strike was going. We had started out with $16 million in our strike fund, but eventually had to spend every penny of it. There was some discussion within the union about whether we should go into debt to continue to fund the strike. I felt that as long as members were willing to fight, even if we had to take out loans to support them—we would. And we did, for an additional $16 million.

However, we found it difficult to get decent terms from U.S. lenders. Without our asking, two officers of the Japanese Telephone workers arrived at CWA headquarters on December 7, 1989—Pearl Harbor Day. The next day we sat together in my office and they took out papers and gave us a $16 million loan, secured only by my and CWA Secretary-Treasurer Jim Booe's signatures. The terms were so favorable that the effective interest we paid came to 2 percent for the three-year loan.

A great trade union leader whom I admired tremendously was Paul Hall, president of the Seafarers Union. Paul always said that a friend is someone who finds out you have a problem and before you even ask for help, is there for you. The Japanese were and still are friends. They also displayed the truest sense of international trade union solidarity.

The support of our friends all around the world greatly contributed to the successful conclusion of the strike. When it ended, a member of NYNEX management said that before we have another labor dispute, they were going to take into account CWA's international clout.

As a result of our long-running fight with Sprint over the firing of 175 La Conexion Familiar workers, we have been making the case against Sprint to our international colleagues and through them, to political figures, around the world. At one point Sprint bought an ad in a London newspaper to attack us, which proved we were getting to them. When Sprint tried to gain a foothold in the United Kingdom, Labor members of Parliament protested, asking whether an American company that exploits women by paying substandard wages and benefits should be welcomed in the U.K. Our German colleagues came to Kansas City to meet with Sprint management about the issue, because they sit on the board in Germany presiding over the joint venture between Deutsche Telecom, France Telecom and Sprint, Global One.

We have also helped our brothers and sisters in other countries. When Bell Atlantic and Ameritech purchased the telephone company in New Zealand, we worked very hard to protect the rights of the New Zealand workers. We had mobilization campaigns both here at home and in New Zealand, and loaned them a considerable amount of money, some of which we never got back when the union was busted because of atrocious antiunion legislation.

During the horrible years of apartheid, when Nelson Mandela languished in jail, we helped lead the South African boycott, particularly against Shell Oil here in America. CWA members marched together at scores of demonstrations. Our union closely supported the Post Office and Telephone Workers Union in South Africa. Among other things, we donated a van to enable them to better service their members. And when apartheid was finally dismantled, several CWA staff, led by Vice President Brooks Sunkett, went to South Africa to be monitors and observers of the election of President Nelson Mandela. I was particularly honored to be in South Africa in 1996 when the three unions—Black, Colored and Indian—that represented the telephone workers, dissolved and formed a single, united union, the Communications Workers Union (CWU). I was privileged to meet with the new officers and pledged our support.

Since 1961 CWA has been supporting the Cuban telephone workers in exile. During the revolution, we helped get the leaders of the Cuban Telephone Workers Union who were targeted by Castro out of the country and safely to Miami. Now their sons and daughters are anxious to return to Cuba and bring the union with them. In 1997 the Communications International contacted the Cuban Telephone Workers and said that we would like to meet with them. But the Cuban union set such onerous ground rules that we could not accept them. For example, we could only meet with whomever they designated. One of these days, we will be able to help bring a free trade union back to Cuba.

In 1985, following our election to office, Secretary-Treasurer Jim Booe and I were trying to come up with an appropriate tribute to the life and work of our retired president, Glenn Watts.

We learned that the Jerusalem Labor Council of Histadrut,

the Israeli Federation of Labor, owned a piece of vacant land in East Jerusalem. Discussions with our colleagues in Histadrut led us to undertake building the Glenn Watts Cultural Center, a building that would, among other things, house the Arab Department of the union.

The property on which the center is built was known as "No Man's Land"—right on the border between East and West Jerusalem prior to the 1967 War. Israelis and Arabs took turns shooting at each other from both sides of the street surrounding the block. Across the street from the center is a building where Jewish sharpshooters once fought. The building is a museum now.

When we held the ground-breaking ceremonies in 1985 for the Glenn Watts Center, the undeveloped property was pockmarked with old shell craters and bullet holes were visible on the few building walls that were still standing. Now it is inspiring to walk through the halls and see young Arab women, daughters of union members, being trained in all aspects of the textile industry. The center has had a profound impact on the life of the Arab community. We are very proud of the role we have played to improve the lives and job skills of Arab youngsters.

In 1998, I was honored when Israel Bonds announced they would fund a computer lab at the Watts Center that would carry my name.

The Glenn Watts building houses the first Arab senior citizen club in history. When the seniors are not using the room, it is where the youth group meets. In 1996, my granddaughter Nikki and a young Arab boy jointly cut the ribbon inaugurating the youth group meeting room.

Today, Jewish youngsters in the neighborhood can be seen playing basketball at the building playground along with their

Arab neighbors. (Some of them are still wearing the "Clinton-Gore" t-shirts we gave them.) In some small way, we may have contributed to peace while honoring a great humanist, Glenn Watts.

The Glenn Watts Center was built with all private financing as a union-to-union project. We received funds from unions and private sources all over the world. By doing something for the Arabs, we wanted to make a contribution to better understanding, never dreaming, of course, that a breakthrough would occur in our lifetimes. We need greater international understanding so that workers of different countries realize that they have more in common than they think.

One of my first visits to Japan was in 1980, when Glenn Watts asked me to speak at a conference in the union's Kinki Region, Osaka. Florence came with me. Our Japanese hosts gave us two separate itineraries. While I went to the executive board meeting, Japanese women took Florence around. At the board meeting, I received a very warm welcome. I thanked them, and then went on to say that I felt at home, since this executive board was very similar to CWA's. "We, too, have only one female," I said. (This was true of CWA in 1980.)

The only person to applaud my comment was the lone female on the board.

I said our one woman had been a telephone operator. Again, one person applauded.

I told them about how affirmative action was changing the role of women in the telephone industry, but made it sound like there was a woman working on every pole in the U.S. and down all the manholes. Of course, while great progress had been made here in America, this was a large exaggeration.

The president said that Japanese women did not want to do

that kind of work. I told him: "You know, we said the same thing until we gave them the opportunity, and learned they wanted to do it."

Once again, the lone woman on the board applauded. I decided to abandon this line of discussion, as my host was clearly irritated.

My wife made quite an impression on the Japanese. On the first day, our hosts had hung a sign in the union office which read: "Welcome, Mr. Morton Bahr, President of CWA." On the second day, after they had got to know her, someone had painted in, "And Florence."

I did not realize that I was the first CWA leader to visit Japan with his wife. Our last meeting on that trip was a lunch with the Japanese representative of the PTTI.

"You don't know what your bringing Florence means to us." Mr. Hatsuoka said. "You'll get home, go about your business and eventually forget about this trip. But Florence will always be an ambassador for Japan."

In 1986, a year after I was elected president, Florence and I led a CWA delegation to Japan. CWA and Zendentsu (the Japanese Telephone Workers Union) had agreed to hold bilateral talks on matters of mutual interest. We would alternate countries every year. At the farewell reception, the three top officers of the Japanese union brought their wives. During my remarks, I said, "Back in 1980, you told me how Florence would be an ambassador for Japan, and six years later it's obvious she really has been. But the United States has so many problems that we really need ambassadors all around the world. So when you come to the United States next year, I hope you will bring your wives so they can be ambassadors for the United States."

You could have heard a pin drop. Then the president's wife

started clapping, and the other two wives followed. From that point on, whenever these forward-thinking Japanese telephone union leaders visit America, they bring their wives.

Florence and I had developed a very close relationship with Akira Yamagishi, the head of the union. Whenever we came to Japan, he brought his wife and two granddaughters to meet us at the airport. He was a man of great insight, progressive, tough and knowledgeable. He was a great leader. It was no surprise, therefore, that when the Japanese labor movement unified and formed RENGO (similar to our AFL-CIO), Yamagishi was elected president.

On one visit, the CWA group wanted to visit Hiroshima. President Yamagishi sensed the tension among our group. He said, "We started it—You finished it." That put all at ease.

Yamagishi said that he was glad the war ended, because he was being trained to be a Kamikaze pilot at the age of sixteen.

"Why in the world would you volunteer for something like that?" I asked.

"You've seen too many American movies." Yamagishi said. "You didn't volunteer to be a Kamikaze pilot."

Our relationship with the Japanese trade union leaders has been a great learning experience for both sides. Japan is beginning to change from its long tradition of being a patriarchal society. We are beginning to see changes, including more female leaders in Japanese politics and even at the lower echelon of the union.

I was once sitting alone in a restaurant when two female economists who worked for the union recognized me. In what still is considered being quite forward, they motioned for me to come over. They introduced themselves and said they loved when I visited because for a while after my visit their conditions of respect improved. So, while change is taking place, it will not be dramatic, but slow and evolutionary.

Being exposed to different cultures expands one's vision and makes it easier to do the job at home. Instead of fitting your mind into a neat little cubbyhole, you are challenged to think outside of your own culture and experience.

In the years ahead we will see more cross-border organizing. While labor laws prohibit CWA from organizing in any foreign country other than Canada, we will continue to support our brothers and sisters in our related fields around the world. We are as committed to helping organize AT&T in Europe as we are here in the United States. Our members are beginning to realize that wherever their employer expands, be it here in America or overseas, if those new jobs are nonunion, it will threaten them. If their employer has nonunion workers, there will be pressure on their jobs, wages and benefits.

As companies expand their overseas markets and operations, many will have greater revenues from outside their home borders. Loyalty will no longer be a consideration for most corporations. If it is, their loyalty will be to the dollar, the yen, or the Deutsche mark. Only the international trade union movement working together, on behalf of the workers of all nations, can exert a stabilizing influence on large multinational corporations. The more workers are organized all over the world, the more secure we all are.

The countries that once belonged to the Soviet bloc and are now emerging into the global marketplace present special challenges to the international trade union movement. These countries have trade unions. Those unions were all part of the communist power structure, and in a few cases are still not completely independent. Before we accept unions from the former Soviet bloc into our international trade secretariat, we send a delegation to meet with them. Our delegation investigates their status and reports on whether they should be invited to join. The sole criterion for membership in the secretariat is whether they are sufficiently independent of the government, even though the governments are now democratic. In the case of Russia, we have still not accepted their application because we have determined that they are not free from government control. However, virtually all of the former Soviet bloc countries are now members of CI.

CWA and the rest of the labor movement fought hard against the NAFTA treaty, not because we are against foreign trade, but because we thought it was a bad policy for America and for Mexico. And now that wages in Mexico are lower today than they were prior to NAFTA's passage, we can say that we were right.

At the same time we fought NAFTA, we also worked hard to create a greater understanding between American and Mexican workers. We are supporting a new trade union movement in Mexico, the UNT. One of the top leaders is the president of STRM, the Mexican Telephone Workers Union, Francisco Hernandez Juarez. He has shown extraordinary courage in opposing the old guard, corrupt trade center, CTM, and its ties to the government in power.

Another area where labor's motives are sometimes challenged is immigration. Our critics seem to forget that the modern American labor movement was founded by immigrants.

U.S. immigration laws set an annual cap on how many foreign "guest" workers may enter the country. In 1998, that cap was hit early as many companies brought into the country workers with varied computer backgrounds. Companies pressed Congress to raise the cap substantially—and for some years in the future.

As chairman of the Department of Public Employees, I took the lead, along with the AFL-CIO, to work with the Clinton Administration to oppose any such increases in quotas unless they were coupled with broad programs to train American workers for those jobs where shortages of workers are expected. This fight continues with strong support from Sen. Ted Kennedy.

We will continue to work toward a strong labor movement in every country of the world. This will enable workers to prosper in a place they call home.

As sure as day follows night, the time will come when unions and their management counterparts at AT&T, Disney, News Corp and all of the other multinational companies will meet regularly to discuss matters of mutual concern. I would prefer it to happen because management sees a mutual value to the dialogue.

By the turn of the century, there will be a single international labor organization for all telecommunications, postal, information, media and entertainment unions and other service workers. That international organization will be in a stronger position to assist the development of free and democratic unions where they do not yet exist in parts of Asia and Africa.

Of equal importance, the new organization, crossing all borders, must be the leading advocate for universal service. The many areas of zero teledensity—more than two billion people who have never made a phone call—must be given access to advanced communications. Only in this way can the international trade union movement bring the benefits of information technologies to every corner of the globe, and thus the promise of economic development and creation of high-wage, high-skill jobs in every country.

CHAPTER SIXTEEN

A Vision for the Future

I f given the opportunity to address the U.S. Chamber of Commerce or the Business Roundtable today, I would offer them the following dream.

"Imagine that your fondest wish came true and tomorrow, with a wave of a magic wand, you awoke to a world without unions. There is no threat of a union coming into your business. The union that was in your company is now gone. There is no collective bargaining. No strikes. No sick outs. Just docile workers grateful to be employed and completely under your control.

"Sound like a fantasy come true? Be careful of what you wish for, because it might happen. Like most dreams, the reality might not live up to the fantasy."

What do I mean?

The yearning of working people to join together to improve their economic lot is as old as the building of the Pyramids in Egypt, where the first recorded strike occurred. Without a trade union movement that rests on the principles of democracy and economic justice, worker organizations will surely rise up that are antithetical to capitalism and free market economies.

History teaches us over and over again that free trade unions are the bulwark of democratic societies. When despots and tyrants seize power, their first objective is to silence and destroy free trade unions. Communist dictators sought to replace free trade unions with their own version of "company dominated unions" which were controlled by the states. Cuba is still an excellent example of this.

Along with several other CWA officers, I had the honor, in Hong Kong, to have dinner with Han Dan Fong, a hero of Tiananmen Square. Han was a twenty-four year old railroad worker who got caught up in the democracy movement and began the Workers Autonomous Union. He told us how he was arrested and tortured; how the Chinese authorities infected him with tuberculosis. But he would never give up. He was and still is determined to bring democracy and a democratic trade union movement to his country. That is when he will go home along with his wife and new baby. Meeting with Han, I felt I was in the presence of a giant. I was not surprised to later learn that he was one of the speakers at the demonstrations in Hong Kong which marked the tenth anniversary of the Tiananmen Square massacre.

In 1981 Ronald Reagan became the first U.S. President to ever destroy a free trade union when he broke the strike of the air traffic controllers by bringing in replacement workers. Other presidents may have helped jail union members and break strikes. President Reagan was the first to actually do away with a union, the Professional Air Traffic Controllers Organization (PATCO), as an institution. But not even President Reagan could extinguish the fires of workers who know that only through organization can they seek to insure justice on the job. Out of the ashes of PATCO came a new union, the National Air Traffic Controllers Association, which continues to represent our nation's air traffic controllers and is now an affiliate of the AFL-CIO.

Two presidents and two decades have passed since President

Reagan fired the air traffic controllers. And today, the union is still there fighting for its members.

As I look back on CWA's history, the men and women of this union have participated in the struggles, challenges, attacks, victories and defeats that have marked the history of organized labor. They have built a union of formidable proportions. They have played a significant role in the creation of America's free trade union movement. And today they are among the leaders in labor's resurgence.

I am deeply honored to have the opportunity to serve them and their families. It is a great privilege to be a union officer in CWA. Union leaders and activists are different from other people. They are not motivated by profit, making a million dollars or advancing to the top of the corporate ladder.

Unlike CEOs, they are elected by the people they serve and they can be unelected if they don't respond to their needs. It happens all the time in CWA. The hallmark of CWA leadership is service to our members, our community and our nation. I am proud to be their leader.

We stand on the brink of a new century and a new millennium. My view is shaped by my life in the 20th century, but my vision sees the new world that our members and their families are entering in the 21st century. I return to the choice before us that I expressed at the beginning of this book: Crisis or opportunity?

Many scenarios lie before us.

The possibility certainly exists that organized labor could become marginalized to the edges of American political, economic and social life. It is not realistic to believe we would ever go out of business. Unions have too much financial strength with pension monies and the responsibility for other funds to simply disappear. Union members also are locked into key industries such as construction, auto, steel and telecommunications, which means that organized labor will never be completely ignored.

But the trade union movement could be seriously weakened through the political and legislative process. "That which is won at the bargaining table can always be lost at the ballot box," is an old labor adage with much truth, particularly today. Just as likely, however, is the ongoing loss of union membership, both in real numbers and as a percentage of the nation's workforce. That is the biggest threat facing the union movement today. Unrestrained employer resistance and the failings of labor law are the two major challenges to union organizing today. We simply must develop new organizing strategies and tactics.

When union labor is strong, the gains in our collective bargaining agreements spread to the nonunion sector, even if for no other reason than the employer wants to keep the union out. When labor is weak, employers feel no incentive to share the wealth with nonunion workers. If labor becomes irrelevant, we will feel the downward pressures of nonunion wages and benefits. Even today when we bargain, management often cites the competitive threat of nonunion companies as a wedge to try and wring concessions from us or drive down our expectations.

With all these seemingly hostile forces bearing down on us, what are the opportunities facing labor and CWA in the future?

AFL-CIO President George Meany used to say that if you don't have 100,000 members in your union, look for a partner, because you cannot serve your members and the labor movement. This was some twenty-five years ago.

Today, in order to properly serve your members and have resources to grow, it is more like 500,000 members.

Smaller unions should view the occasion of a new century as a time to seriously reassess their viability in the face of ongoing corporate mergers, technological advancement, political attacks and global competition. Labor must concentrate its financial and membership resources to effectively respond to these pressures.

A smaller union, even one with a proud history and active membership, is just no match for a multi-billion dollar corporation that has the latest technology at its fingertips and can transfer money, jobs and information anywhere in the world with the push of a button.

CWA has been very successful in merging with smaller unions in a way that preserves their identity and heritage. We see our mergers as partnerships that are reshaping CWA into a union that is far different from the one that I joined in 1954 or first became president of in 1985. Smaller unions need not be lost in a merger with a larger union, nor should their members feel threatened with being ignored. We are actively engaged in discussions with small and large unions about how our organizations can join together to form a bigger, stronger and more effective voice for working families. Our goal is always to combine the best of both organizations.

When we do complete a merger partnership with another union, we also enlarge the future leadership pool of our union. Union leadership development is another of the critical challenges I see for labor in the future. Where will the George Meanys, Walter Reuthers, Joe Beirnes, A. Philip Randolphs, Al Shankers and John Sturdivants of tomorrow come from? Younger workers, women and minorities must not be simply encouraged to stand for leadership positions, they must be actively recruited. We can only do that by insuring diversity in our ranks and our leadership. This is not being politically correct. Diversity in our unions is an essential investment in the future leadership of our movement.

For all these reasons, union mergers must be seen as part of the overall process of labor redefining itself for the next century. There are just too many unions, most of which are unable in terms of finances, resources or membership to withstand the incredible pressures that we all face today. That's why I applaud the leaders and members of the Steelworkers, Autoworkers and Machinists for their visionary approach in building a new metal trades union. They have inspired all of labor to take a deep, hard look at itself. Their new partnership will not be easy, but I know they will move forward to overcome the potential obstacles. The stakes are too high for them not to be successful in this historic endeavor.

The potential merger between the American Federation of Teachers and the National Education Association is another incredible milestone. Again, these two great unions are seeking to form a new organization that will be a powerful voice for students, teachers and other professionals. If they do, the national debate and politics of public education will never be the same. CWA offers its encouragement and support to the leaders and members of these two unions to stay on course for the good of all union members and our nation.

Every labor organization will be buffeted by enormous change in the next millennium. To be successful, it will take an acute understanding of how change will impact the union itself, as well as how the union recognizes what the changes themselves mean. At the turn of the 20th century, the horseshoers union clung to tradition and missed the opportunity to transform itself into the Autoworkers union. It took labor nearly forty years to organize a dominant world industry only after great struggle and sacrifice. With a little foresight, auto could have been a unionized industry literally at the ground floor.

CWA does not intend to make that mistake as we enter the 21st century. As a union that has always stood at the doorstep of technological advancement throughout its history, we have demanded from our employers union recognition for all of the jobs impacted and created by new technology.

We will also have to undertake a serious re-examination of ourselves to see how we can change to better meet the needs of professionals. We will explore new technologies that may allow us to communicate directly with the membership and them with us. On the educational front, we should investigate CWA-sponsored, accredited certification and degree courses that can be brought into the workplace or that our members can take online. We should assume a greater responsibility for serving as a focal point for networking and career opportunities for our members. Professional recognition, travel that enhances career experience, professional conferences and meetings are just a few of the new benefits CWA could offer to our members in the professional, technical and administrative occupations.

In other words, we should look at every opportunity to create a new kind of workers' organization that combines the power of a labor union on the job with the professional and social enrichment of an association.

Another target of opportunity for labor is international trade. Organized labor achieved a major breakthrough when Congress rejected President Clinton's request for fast track trade negotiation authority. This was the first time a president had been denied fast track authority by Congress and the first time that labor successfully injected human and environmental rights into the trade debate.

Our victory on fast track opened many people's eyes to the threat of unrestricted free trade to workers' rights and to the environment. Our lawmakers finally had the courage to stand up and say, child labor is wrong; exploitative labor is wrong; prison labor is wrong; and trashing the environment for profit is wrong.

This will help to shed the "protectionist" label that is always used to describe labor's opposition to unrestricted free trade and engage the nations of the world in a serious debate over appropriate international labor and environmental standards that should govern our trade relations.

The modern U.S. labor movement has never opposed open and fair trade between nations. We ask only that international trade lift the living standards of all workers and that competition be based on product innovation, quality and productivity, not just which nation can pay the cheapest wages. We cannot see meeting these goals with our current trade policies that tend to ignore human suffering and environmental degradation. We have lost too many U.S. jobs and seen too many U.S.-based industries decimated to cheap-wage nations that fail to enact or enforce basic laws protecting workers rights and the environment.

Those who continue to defend the current system in the face of the negative experience of millions of American workers risk undermining their own credibility and the people's support for international trade. That's just what happened in the fast track debate. Labor must continue to build on the public support that we have mustered to force world leaders to pay attention to the concerns of workers.

Finally, organized labor has the opportunity to take the national lead in rekindling the feeling of community and caring for one another that is the true spirit of America. Mean-spiritedness and nastiness have overtaken our nation's life for too long. We live in an era when prominent business executives publicly attack union leaders such as myself as "thugs" and "threats" to America. In the name of entertainment, radio talk show hosts describe to their listeners the most effective way to shoot federal law enforcement agents. Selfish individualism has replaced community service. Civil discourse in our national discussions has been taken over by personal attacks and name calling. The chairman of a congressional committee called the president of the United States a "scumbag."

Organized labor has a unique opportunity to issue a national call of union and community service to our members. Such action would be a win-win both for our nation and for our movement.

There are approximately 16 million union members in the United States. Each of them can have a profound impact on making the dreams that most workers share become a reality.

I recognize that all 16 million are not committed trade unionists. Notwithstanding, they can make a difference. The union member(s) in each family must make union and union-related matters subject for discussion at the family dinner table or as opportunities arise.

We are all affected by our environment. My family certainly was. As I look back, it is very obvious that growing up in a union environment has made the entire Bahr family more caring members of the community, as well as being a factor in their selection of careers.

My wife Florence was most active in our community in various ways, including cancer research, PTA and More Beautiful

Port Washington. She served in several capacities in the Jewish women's organization, Hadassah, and then as president of the local chapter. My daughter Janice was also an active member in many different organizations and was elected president of her Hadassah chapter in another city.

I remember getting a call from my son Dan. He was a speech therapist in a Long Island school district. He told me how lousy conditions were. I suggested he do something about it. A year later he called and reminded me of the conversation, and that he had done something about it. I asked, "What?"

"I was elected local president," he replied.

After accepting my congratulations, just like I had exclaimed so many years earlier, he asked: "What do I do now?"

Dan went on to become a member of the staff of the New York State United Teachers (American Federation of Teachers). In 1984 he took a leave of absence to become the director of employee relations for Suffolk County. In that role he had the full confidence of the County Administration and the twelve unions that represented the employees.

Dan later returned to his first love, his job at the Teachers Union. I always have a great sense of pride when I am told by those whom he works for or with about the great job he does.

Dan's oldest daughter Heather, a graduate of SUNY-Albany, works for Suffolk County as a "labor technician." She is involved with job training and the welfare-to-work program. I am absolutely certain that she gravitated to this kind of work because of her exposure to CWA conventions, meetings, conversations, and the surroundings at her grandparents' home and, of course, growing up in a second-generation union family.

As a third grader, my youngest granddaughter, Alison, called me excitedly to tell me she won the first prize among all the third graders for the best poster depicting "What liberty means to me." Her poster featured women pickets protesting. Alison has also written about Pediatric AIDS. I was so proud of what she wrote that I sent it to Elizabeth Glaser.

All four of my granddaughters were at the convention in 1985 when I was elected president. The New York locals held a reception one evening. Florence brought the girls. Shelley, ten years old, disappeared for a few minutes. Florence scolded her and reminded her that she was not supposed to be out of sight. Shelley remarked: "Grandma, there are no perverts in CWA."

Granddaughter Nikki, along with Shelley, travelled with us to union meetings in various parts of the world. Nikki is in a masters program and will teach upon graduation in September 1999.

I don't believe these were accidents. All of them are sympathetic to the union movement; none would ever cross a picket line. More importantly, perhaps, each is a warmer, more caring person because they share the values of trade unionism.

As trade unionists, we owe it to the next generation to make the home and the family the place where basic human and union values are taught.

For some sixty years, CWA members have actively practiced these values. When our union first began, there was still a depression. The world stood on the cliff of an unimaginable holocaust. The atom bomb hadn't yet been invented. The only cold war we fought was in winter. There were no televisions, computers or electronic games. A grade school education was sufficient to get a job.

CWA now embarks on our next sixty years. The Cold War is over. Four-year olds play on personal computers and navigate the Internet. The global marketplace is a reality and today's employer is just as likely to be headquartered in Brussels as in Chicago. Lifelong education is now a necessity for workers just to remain employable. And a new drug promises to restore the virility of men over fifty.

What a time.

But the principles which sustained us in the past will most certainly succor us in the future. Where CWA was conceived as a union for telephone workers, we now are poised to rise in the next century as the union for professional, technical and administrative workers in all walks of life in both the public and private sectors.

As we look to the future, I am reminded of the words from the song "Pass It On" which was the theme song from "The Inheritance," a moving film produced by the former American Clothing Workers of America (later named the Amalgamated Clothing and Textile Workers and now UNITE).

Freedom doesn't come like a bird on the wing,
    doesn't come down like summer rain.
Freedom, freedom is a hard won thing,
    you've got to work for it!
Fight for it!
Day and night for it,
    and every generation's got to win it again.

Pass It On to your children, mother.
Pass It On to your children, brother.
You've got to work for it, fight for it, day and night for it.
Pass It On to your children,
Pass It On!

CWA is the voice for the workers in the information age. We are the leaders in bringing the benefits of the information revolution to all of our citizens and to people around the world. We are the union for the future.

We now embark together on a great adventure in a new century. I am confident we will become a stronger union and contribute to making the United States a better nation. Together we need to make the new technology work for all peoples of the world, lifting them out of poverty and making the American Dream a possibility for all who live on this planet.

Uncertainty lies ahead. There will be bumps in the road. Our struggle may be never ending. But what an exciting future we have.

I can't wait.

CWA History— A Chronology

1910-1919 EARLY ORGANIZING EFFORTS IN THE TELEPHONE INDUSTRY.

The first union to attempt to organize telephone workers—the International Brotherhood of Electrical Workers (IBEW)—achieved limited success during these years. It was not until 1912 that the IBEW accepted telephone operators—generally women—as members. In 1919, IBEW's telephone department claimed 200 telephone locals with 20,000 members.

1918-1923 WORLD WAR I—THE GOVERNMENT TAKES CONTROL OF THE TELEPHONE SYSTEM

Under a presidential order on July 22, 1918, the telephone and telegraph system was placed under the control of the federal government and Postmaster General Albert S. Burelson. In 1919, Burelson was faced with a strike by the IBEW that virtually tied up phone service in New England and threatened to become nationwide. In an attempt to end the strike, Burelson issued a government bulletin acknowledging the right of workers to bargain through committees "chosen by them, to act for them."

1920-1935 GROWTH OF COMPANY UNIONS IN THE TELEPHONE COMPANIES

Frightened by the prospect of legitimate unionism on a large scale, AT&T encouraged employees to form and join company dominated unions (usually called associations or committees). The company unions succeeded in virtually destroying the existing IBEW telephone locals. By 1923, IBEW had been ousted in every location except Montana and the Chicago Plant. Company unions dominated the telephone industry until 1935.

1935   CONGRESS DECLARES COMPANY UNIONS ILLEGAL

Congress passed the National Labor Relations Act (more commonly known as The Wagner Act) which prohibited employers from engaging in unfair labor practices, such as establishing company unions. It protected union activities such as grievances, on the job protests, picketing and strikes. And it established an agency, the National Labor Relations Board (NLRB), to enforce the new labor law.

1937   SUPREME COURT DECLARES NATIONAL LABOR
             RELATIONS ACT CONSTITUTIONAL

Upon the Supreme Court's decision that the NLRA was indeed constitutional, company unions were no longer an option in the telephone and other industries.

1938   GROWTH OF INDEPENDENT TELEPHONE UNIONS AND
             THE CREATION OF NFTW

After preliminary meetings in St. Louis and Chicago, representatives of thirty-one telephone organizations, representing a total combined membership of 145,000, assembled in New Orleans in November 1938, and adopted a constitution and established the National Federation of Telephone Workers (NFTW).

NFTW was a federation of sovereign local independent unions and its lack of authority over the affiliated local unions left the NFTW at a serious disadvantage in dealing with the Bell monopoly.

1941-1946 WORLD WAR II AND THE NATIONAL WAR LABOR BOARD

Following the attack on Pearl Harbor, the AFL and CIO voluntarily gave no-strike pledges to the federal government for the duration of the war. In January 1942, President Roosevelt created the National War Labor Board (NWLB), which was charged with settling all disputes between labor and management that threatened war production.

The NWLB and twelve Regional War Labor Boards were composed of an equal number of representatives from management, labor and the private sector. All of the labor representatives appointed to the Board came out of the AFL or CIO. This was a great concern to the NFTW, which was not affiliated with either organization.

The average real wage of a telephone worker dropped from 83 cents an hour in 1939 to 70 cents an hour in 1943. In 1939, telephone workers occupied the twenty-second place on a list of average weekly earnings of workers in 123 industries. By early 1945, they had fallen to eighty-sixth place on the list.

Telephone unions brought numerous cases before the NWLB and the Regional Boards. Response to these appeals was exceedingly slow, and by mid-1944 there were eighty-five cases brought by telephone unions still waiting to be ruled upon.

1943   FIRST BLACK OPERATOR HIRED IN NJ BELL SYSTEM

Gloria Shepperson was the first black operator to be hired in the NJ Bell System—most likely in the entire Bell System. Shepperson had to fight an anti-discrimination case to win her job as an operator. She was appointed to the CWA staff in District One by Vice President Morton Bahr and went on to become CWA's Director of Ethnic Affairs and Assistant to CWA Secretary-Treasurer Louis Knecht.

1944   DAYTON, OHIO STRIKE AND ESTABLISHMENT OF THE NATIONAL TELEPHONE PANEL

In November 1944, Dayton telephone workers went out on strike and within six days, the strike spread to twenty-five cities in Ohio, Washington, D.C., Chicago and Detroit. At that point, the government capitulated and agreed to establish a national board modeled on the NWLB that would only handle the cases of telephone workers.

On December 29, 1944, the National Telephone Panel (later to be renamed the National Telephone Commission) was established. With two members each from the public, industry and telephone labor sectors, its mandate was to hear and adjudicate all telephone cases and to formulate basic telephone wage policy.

The Telephone Panel was much more effective than the NWLB. By the end of 1945, when it was terminated, it had heard fifty-five disputes involving 180,000 workers.

1946   FIRST NATIONAL AT&T AGREEMENT

When the war ended in August 1945, the wages of telephone workers remained below those of many industries. Contract negotiations stalled and the presidents of the NFTW affiliates authorized the union's Executive Board to call a nationwide strike at 6:00 a.m., March 7, 1946. In the early morning hours of March 7, workers around the country prepared to walk the picket lines.

At 5:30 a.m., after twenty hours of bargaining, NFTW President Joseph Beirne and Cleo Craig, AT&T vice president in charge of negotiations, came to an agreement. A strike had been avoided and for the first time in history, AT&T had negotiated a national agreement with the union and committed its associated companies to that agreement.

While a major victory was won in the 1946 negotiations, the basic weakness of the NFTW had been revealed. During negotiations, 34 of 51 affiliated unions broke away and signed separate agreements.

1947   THE STRIKE THAT BROUGHT AN END TO THE NFTW

In 1946, AT&T was not prepared for a strike. But in 1947, AT&T was not only prepared for a strike, it forced NFTW into strike action. AT&T was determined not to repeat the Beirne-Craig type of national settlement. It flatly refused to bargain on an industry-wide basis. The company did not make a wage offer until three weeks into the strike and made the offer contingent upon the affiliates agreeing not to clear it with NFTW's policy committee. Five weeks after the strike began, seventeen contracts had been signed. The strike collapsed and the NFTW was finished.

NFTW President Beirne summed up events by saying: "We were trying to make a federation of unions do the job which can only be done by one union in the telephone industry."

During the 1947 strike, AFL and CIO unions lent their moral and financial support despite the fact that NFTW was not affiliated with either at the time. International unions in both the AFL and CIO aided the strikers with contributions totalling $128,000. This support was very important in helping NFTW workers survive the strike and regroup into a strong and truly national union.

1947   THE FOUNDING OF CWA

In June 1947, the Communications Workers of America, came into being. The first CWA convention took place that month in Miami, with 200 delegates representing 162,000 workers.

The delegates adopted the first CWA constitution. Joseph Beirne was elected president and advised the delegates: "We must embrace the 'all for one and one for all' philosophy of a single CWA union."

1949   CWA AFFILIATES WITH THE CIO

In February 1949, CWA's Executive Board recommended affiliation with the Congress of Industrial Organizations (CIO) and in a referendum, the membership approved the CIO affiliation. Not all of the independents agreed to join CWA. The AT&T Long Lines unit applied for a CIO chapter which was granted as the Telephone Workers Organizing Committee-CIO (TWOC-CIO). Later, units from AT&T manufacturing and sales and Michigan Traffic joined the TWOC-CIO. When CWA was granted the CIO charter, TWOC was folded into CWA.

1950   U.S. SENATE CONDEMNS BELL SYSTEM

In response to charges levied by CWA, the Senate Subcommittee on Labor-Management Relations held hearings to investigate the status of collective bargaining and labor-management relations in the Bell System.

Following the conclusion of the hearings, a majority report of the subcommittee overwhelmingly supported the charges made by CWA. The subcommittee found that:

1. The local associated companies functioned as parts in a closely integrated corporate system completely and directly controlled by AT&T management;

2. The basic cause of poor labor-management relations in the Bell System revolved around the inability of the union to bargain at a level of management that had the authority to make final decisions; and,

3. The Bell System had actively and continuously conducted an antiunion campaign, including ads in the public press and interference in CWA affairs.

1951   CREATION OF A NATIONAL DEFENSE FUND

In 1951, after two days of heated debate on the issue and a roll call vote with 133,047 in favor and 101,883 opposed, the delegates to the annual convention voted to establish a national defense fund with contributions of 50 cents per member per month.

1955   SOUTHERN BELL STRIKE

CWA undertook its most difficult challenge in its short history: a regional strike against Southern Bell lasting seventy-two days, encompassing nine states and affecting 50,000 workers. Throughout months of bargaining the company remained adamant that any new contract contain a ban on strikes "or other interruptions of service." Throughout the strike, CWA expressed its willingness to resolve bargaining issues through arbitration, but Southern Bell refused.

Ultimately, Southern Bell's attempt to break the union failed. A one-year contract was signed that gave across-the-board gains to CWA members: wage increases, the right to arbitration for suspensions, discharges and job vacancy fillings, reduction of work tour hours, and most significantly, recognition of the right to strike. The 1955 strike was an early landmark for CWA because of its scope, duration and success.

1963   General Telephone of California Workers Demand Equal Pay for Equal Work

In October 1963, CWA members went on strike against General Telephone of California for wages and benefits comparable to those enjoyed by Bell employees in the state. At the time, it was possible for a General Telephone worker and a Bell worker to be doing the same type of work across the street from each other, but the General Telephone employee would be receiving considerably less compensation for the job than his/her Bell counterpart.

1965   The Triple Threat Program—Organizing Growth Resolution #1

Convention delegates, at President Beirne's urging, adopted CWA Growth Resolution #1, which endorsed the Triple Threat program and clearly stated that organizing was a top priority of the union. It was Beirne's program for broadening the membership base and expanding CWA's influence in the areas of politics and education as well as collective bargaining.

1966   Public Worker Organizing

The 5,000 member Municipal Management Society came into CWA and became Local 1180. Between 1966 and 1980, CWA organized the parking enforcement agents and the Board of Elections workers in New York City as well as 15,000 welfare, city and county workers in New Jersey.

1968   National Strike

In the first national strike against the Bell System since 1947, some 200,000 CWA telephone workers walked out because AT&T refused to agree to wage increases that would meet the rise in the cost-of-living. The strike lasted eighteen days with AT&T ultimately agreeing to a raise in wages and benefits totalling nearly 20 percent over a three-year period.

1970   GOVERNMENT CHARGES AT&T WITH
DISCRIMINATORY EMPLOYMENT PRACTICES

On December 10, 1970, the U.S. Equal Employment Opportunity Commission (EEOC) filed charges against AT&T and its twenty-four operating companies for discriminating on the basis of sex, race and national origin in their employment practices.

On January 18, 1973, AT&T, the EEOC, the Department of Labor and the Justice Department agreed on a consent decree, providing for compensation for the victims of past discrimination and an affirmative action program for changing the pattern of discrimination in the Bell System. The settlement included $15 million in back pay to 13,000 women and minority men, and an estimated $30 million in wage adjustments for women and minority workers. A second consent decree signed on May 30, 1974, provided $30 million back pay and wage adjustments to 25,000 employees in lower management positions.

1971   BIGGEST SETTLEMENT IN CWA HISTORY—TASK FORCE
'71 RECEIVES CREDIT

Four hundred thousand CWA members nationwide went on strike against the Bell System in 1971 for wage increases to offset the devastating inflation of the previous three years. After a one-week strike, CWA achieved the biggest economic package ever negotiated with the Bell System and obtained, for the first time, a cost-of-living adjustment clause (COLA) and big city allowance.

While the 1971 strike lasted one week nationally—for 37,000 New York Telephone plant workers it lasted 218 days. This unit achieved a breakthrough in union security by obtaining an agency shop which extended to the entire Bell System in 1974.

1971   FIRST SPECIAL CONVENTION OUTLINES DUTIES AND
RESPONSIBILITIES OF LOCALS

More than 1,500 delegates, alternates and guests attended CWA's first "special convention." The delegates adopted several constitutional amendments, the most important of which required all locals to carry out the union's policies, participate actively in political and legislative activities, participate in local officers' and stewards' training programs, and attend all District, state and local meetings. Another constitutional amendment adopted at the convention created CWA Retired Members Clubs and provided three-year terms of office at both the international and local level (prior to 1971 there were two-year terms).

1973 GENERAL TELEPHONE WORKERS IN THREE STATES WALK OUT

6,000 CWA members in Indiana, Ohio, and Kentucky went on strike against the General Telephone companies. The strike lasted two months in Indiana and Ohio before settlement was reached, but the workers in Kentucky were on the picket line for five months before their contract demands were met. The Kentucky workers were forced out on strike again in 1976. This time the strikers were out for 200 days before a settlement was reached.

1973-1974 CWA DEALS WITH EQUITY AND DISCRIMINATION WITHIN THE UNION; FWTW MERGES WITH CWA

During the 1973 CWA convention, extensive discussions were held on the methods by which CWA dealt with the problems of women and minority members. As a result of these discussions, the National Executive Board established a Blacks and Other Minorities Structure Study Committee and a Female Structure Committee. In November 1973, these committees convened at CWA headquarters and prepared reports for the Executive Board which included recommended policies and procedures.

Extended discussions at the Executive Board meetings in January and February 1974 led to a resolution recommending that the President develop a Committee on Equity concept from the national to the local level of the union. The Executive Board authorized the appointment of a National Committee on Equity consisting of rank and file members from each District, which is still in place today.

The Federation of Women Telephone Workers of Southern California (FWTW) merged with CWA. Its president, Dina Beaumont, became the first female CWA vice president in more than two decades.

1974 FIRST NATIONAL BELL SYSTEM BARGAINING; DEATH OF BEIRNE AND ELECTION OF WATTS

In January 1974, President Beirne left his sick bed to announce to the members of the Collective Bargaining Council that AT&T had agreed to his twenty-eight year old objective—national bargaining.

Beirne did not seek reelection and died on Labor Day, 1974. He was succeeded by Secretary-Treasurer Glenn Watts, who had first gone to work for C&P Telephone in 1941.

1975-1976 STRIKES HIT INDEPENDENTS
Three of the most bitter CWA strikes of the 1970s took place at independent telephone locations: a six-month strike at Rochester (New York) Telephone over an attack on wage levels; at General of Kentucky in 1976 over medical benefits and work rules; and, a three-month walkout at New Jersey Telephone over the issue of supervisors performing bargaining unit work.

1978 FIRST NATIONAL WOMEN'S CONFERENCE
CWA held its first National Women's Conference in Minneapolis, Minnesota. Conference participants attended plenary sessions, workshops and discussion groups. Resolutions on the Equal Rights Amendment, child care and job pressures were presented to the CWA Executive Board.

1979 NATIONAL ORGANIZING DEPARTMENT ESTABLISHED
On July 12, 1979 the Executive Board authorized President Watts to establish CWA's National Organizing Department.

1980-1981 CWA ORGANIZES PUBLIC WORKERS — PUBLIC WORKER DEPARTMENT CREATED
Over the course of the decade, CWA began to expand into fields outside of telecommunications. In July 1980, the CWA Public Workers Department was created. One of the biggest successes in the public sector came in 1981 when 36,000 New Jersey state workers voted to join CWA. CWA now represents some 100,000 public and health care workers.

1981-1983 THE COMMITTEE ON THE FUTURE AND THE SPECIAL CONVENTION
The Committee on the Future was created in July 1981 by action of the CWA Convention. After a year and a half of study and debate, the Committee on the Future submitted its final recommendations to delegates to a special convention in Philadelphia in March 1983. The 1,750 delegates adopted ten resolutions and two constitutional changes proposed by the Committee on the Future.

1982   FIRST NATIONAL CONFERENCE ON MINORITY
CONCERNS

The first National Conference on Minority Concerns was held in Dearborn, Michigan. Participants representing more than 100 locals attended workshops on assertiveness training, leadership skills training for minority workers, effective persuasion through verbal communications, building minority coalitions and coping with stress.

1983   CWA STRIKES THE BELL SYSTEM

Only months before the Bell System was to be broken into separate companies, CWA opened national contract negotiations. On August 7, 700,000 CWA members went on strike for better wages, employment security, pension plan changes and health insurance improvements. The strike lasted twenty-two days when the telephone industry agreed to meet the union's demands. This would be the last time CWA would be able to negotiate at one national table for all its Bell System members because divestiture was only a few months away.

1983   FIRST MINORITY LEADERSHIP INSTITUTE

In response to recommendations by the National Committee on Equity for training opportunities devoted to minorities, the Executive Board established the Minorities Leadership Institute—an annual three-week intensive study program.

1984   DIVESTITURE AND BEYOND

The divestiture of AT&T became reality on January 1, 1984. During this difficult period, President Watts often reminded the members that it was AT&T that had broken up, not CWA—the union remained as unified, committed and strong as ever.

Also in this year, members of the Federation of Telephone Workers of Pennsylvania voted overwhelmingly to merge with CWA. The Executive Board created District 13 to accommodate the 12,250 newly-affiliated men and women.

1985   CWA ELECTS MORTON BAHR AND JAMES BOOE

President Glenn Watts and Secretary-Treasurer Louis Knecht retired after serving eleven years in those offices. Elected to replace them were District 1 Vice President Morton Bahr and Executive Vice President James Booe.

1986   POST-DIVESTITURE BARGAINING; MOBILIZATION TAKES ROOT IN NEW JERSEY

Twelve years after CWA had achieved national bargaining, the union had to bargain not only with AT&T, but with the independent RBOCs and their subsidiaries. National bargaining was replaced by forty-eight different bargaining tables.

In AT&T negotiations, the company attempted to take back health care benefits, lower clerical wages, and eliminate cost-of-living adjustments obtained in earlier contracts. CWA had no choice but to strike. The strike lasted twenty-six days and AT&T agreed to provide wage and employment security improvements and retain health care benefits intact. Although negotiations with the RBOCs were also difficult, they were less contentious than those with AT&T. Strikes were necessary against some of these operating companies, but none lasted more than a few days.

The first large-scale mobilization effort began with New Jersey clerical and professional state workers. Faced with no right to strike in New Jersey, state workers launched the Committee of 1,000 to involve members in mobilization activities aimed at pressuring the employer during bargaining. A strong mobilization system of organization, education and collective action resulted in gaining a breakthrough contract for state workers.

1987   INTERNATIONAL TYPOGRAPHICAL UNION MERGES WITH CWA

Members of the oldest union in the AFL-CIO representing union typesetters and mailers throughout the U.S. and Canada approved affiliation with CWA. Recognizing the distinct nature of the work these members perform, the union created a new Printing, Publishing and Media Workers Sector (PPMWS) and an international PPMW sector vice president.

1988   CELEBRATING FIFTY YEARS OF ACHIEVEMENT; CWA KICKS OFF MOBILIZATION

In 1988 CWA celebrated its 50th Anniversary. The convention took place in New Orleans, the site of the NFTW's founding in 1938. CWA mobilization was kicked off at the 1988 convention in preparation for a major round of bargaining the following year.

1989   MOBILIZATION KEY AT AT&T SETTLEMENT, NYNEX
STRIKE

Mobilization by CWA members around AT&T bargaining caused the company to back off on health care cost-shifting demands. The settlement for 175,000 workers broke new ground on child and elder care by creating a $5 million fund to establish care centers and support facilities, granting parental and elder care leave with a job guarantee and paid medical and dental for six months.

Mobilization also was key for NYNEX workers who spent seventeen weeks on the picket line fighting management attempts to shift health care costs. Local 1103 member Gerry Horgan lost his life on the picket line when he was struck and killed by a scab driving a car at a NYNEX facility.

1991   CWA'S MISSION FOR THE NINETIES: "WALL TO WALL"

Delegates to CWA's 53rd Convention resolved that the 1990s would be the decade of "CWA Wall to Wall." Delegates also made changes to the CWA Constitution to allow the Committee on Equity and the Women's Committee to give annual reports and recommendations to future conventions.

In an effort to put the bitterness of the 1989 strike behind, CWA and NYNEX negotiated an unprecedented early settlement eleven months before contract expiration. The agreement called for a 13 percent wage hike, retention of COLA and fully paid health care. It included a breakthrough agreement on company-wide organizing, neutrality and card check recognition.

1992   MEMBERSHIP INCREASES WITH AFFILIATIONS; CWA
ELECTS FIRST WOMAN SECRETARY-TREASURER

The National Association of Broadcast Employees and Technicians (NABET) affiliated with CWA. CWA became the biggest union in Texas following the affiliation with the Combined Law Enforcement Association of Texas (CLEAT).

Members at AT&T worked for six weeks beyond contract expiration during the summer of 1992, carrying out an extensive mobilization strategy against the company. It was the largest bargaining unit ever to attempt a coordinated inside tactics strategy—100,000 members, five hundred locals, fifty states and thousands of work locations. The coordinated inside tactics by members and massive external mobilization efforts, including generating community and AT&T customer support, proved that sometimes applying pressure in different ways can work better than a strike.

Barbara Easterling was elected as the union's first female Secretary-Treasurer. Easterling had served as executive vice president since 1985 and was a one-time telephone operator.

After more than thirty years headquartered at the Mercury Building in Washington, D.C, the union moved to a new building near the Department of Labor and the U.S. Capitol.

1993   ORGANIZING NEW UNITS, FIGHTING FOR LABOR REFORM

Four thousand graduate students working as teaching and graduate assistants at the State University of New York (SUNY) saw the end to a thirteen-year organizing struggle when they finally voted for union representation. Seventeen hundred clerical and technical workers at the Bloomington campus of Indiana University voted for CWA after a four-year campaign. The Union of Professional and Technical Employees (UPTE) with seven hundred members also affiliated with CWA. The professional and technical workers hold nonacademic positions throughout the nine campus University of California system.

On June 30, CWA activists joined with other trade unionists and community activists in a day of protest at thirty regional offices of the National Labor Relations Board demanding justice from regional directors and labor law reform. The civil disobedience led to 500 arrests.

1995   EASTERLING BREAKS GLASS CEILING AT AFL-CIO; MOBILIZATION MAKES THE DIFFERENCE IN 1995 BELL ATLANTIC BARGAINING; UNIVERSITY RESEARCH PROFESSIONALS AND TECHNICIANS JOIN CWA

CWA's Barbara Easterling made history when she became the first woman in history to hold the position of AFL-CIO Secretary-Treasurer, the Federation's second highest post.

After five months with no contract and an intensive CWA mobilization campaign, Bell Atlantic offered its 37,000 CWA workers double digit wage and pension increases, employment security protections and access to future jobs.

In 1995 and early 1996, 7,800 professional and technical university workers from the University of California joined the CWA family.

1997   TNG JOINS CWA; CWA OBTAINS HISTORIC CARD
CHECK AGREEMENT WITH SBC AND PACTEL;
US AIRWAYS WORKERS WIN A CWA VOICE;
CWA ENDORSES ATLANTIC ALLIANCE.

The Newspaper Guild (TNG), representing news industry workers in the U.S. and Canada, merged with CWA.

A five-year campaign that integrated continuous bargaining, membership education, political action, mobilization and strategic organizing, culminated in March, 1997 when CWA and SBC signed the most far-reaching card check agreement in the union's history. A similar agreement was reached with PacTel in April.

CWA won the biggest private sector organizing victory in a decade when 10,000 passenger service professionals at US Airways voted to join the union.

CWA joined forces with two of Britain's biggest telecommunications unions—the Communications Workers Union and the Society of Telecom Executives—and endorsed an Atlantic Alliance of the three unions to exchange information and plan coordinated strategies to protect our members and to organize new members in the global telecommunications marketplace.

Communications Workers of America, AFL-CIO, CLC

501 Third Street, NW

Washington, DC 20001

(202)434-1100

www.cwa-union.org